DNA Decipher Journal

Volume 8 Issue 3

November 2018

Tooth Regeneration through Wave Genetics, Emotive Source, & Bio-harmony

ISSN: 2159-046X

DNA Decipher Journal
Published by QuantumDream, Inc.

www.dnadecipher.com

Table of Contents

Preliminary Report

Articles

Essay

Exploration

Preliminary Report

A Case Report of Tooth Regeneration in Dog through Wave Genetics

Peter P. Gariaev*, P. P. Vlasov, R. A. Poltavtseva,
L. L. Voloshin & E. A. Leonova-Gariaeva

Institute of Quantum Genetics LLC, Moscow, Russia

ABSTRACT

A preliminary study showing dog tooth regeneration through wave genetics is reported. Regeneration was carried out by a special laser technology based on an expanded understanding of the principles of genetic coding. One of such principles predicts the existence of quantum equivalents of working genes. The multipotent mesenchymal stromal cells ("MMSC") derived from human adipose tissue were pre-treated with human tooth rudiment quantum genetic information (MBER) and transplanted to the area of a test dog where the tooth was removed. The control area where the tooth was also removed was left untreated. After 9 months, a complete regeneration of the tooth was observed in the test dog. In the control area, there was no regeneration.

Keywords: Quantum regeneration, quantum equivalent, genes, stem cells, reprogramming.

The case reported here is devoted to an attempt to regenerate a tooth in a test dig using the method of linguistic-wave genetics [1-2]. The work reported here is preliminary and will be continued. A similar experimental work on the diabetic foot was carried out by us recently [3].

A preliminary study showing dog tooth regeneration through wave genetics is reported. Regeneration was carried out by a special laser technology based on an expanded understanding of the principles of genetic coding. One of such principles predicts the existence of quantum equivalents of working genes. This we proved experimentally in [5]. The multipotent mesenchymal stromal cells ("MMSC") derived from human adipose tissue were pre-treated with human tooth rudiment quantum genetic information (MBER) and transplanted to the area of a test dog where the tooth was removed. The control area where the tooth was also removed was left untreated. After 9 months, a complete regeneration of the tooth was observed in the test dog. In the control area, there was no regeneration.

*Correspondence: Peter Gariaev, Ph.D., Quantum Genetics Institute, Maliy Tishinskiy per. 11/12 - 25, Moscow 123056, Russia.
Email: gariaev@mail.ru

Method & Results

Multipotent mesenchymal stromal cells derived from human adipose tissue were used for transplantation. Cellular suspension from adipose tissue was diluted with Dulbecco's Phosphate Buffered Saline (DPBS), ("Gibco") 1:2, layered on a density gradient of Histopaque 1.077 ("Sigma") and centrifuged for 30 minutes at 600g. Then, interfacial mononuclear rings were collected into centrifugal test tubes ("Corning"), washed by centrifugation in excess DPBS. The resulting cell sediment was resuspended in a culture medium and placed in culture test tubes ("Corning", 25cm^2) and transferred to a 37°C constant incubator with 5% carbon dioxide gas.

The culture medium consisted of DMEM/F_{12} supplemented with 25mM HEPES, 2mM L-glutamine, 2mM sodium pyruvate, 100U/ml penicillin, 100μg/ml streptomycin and 10% fetal bovine serum (all of the above are the reagents of "Gibco"). The medium with non-adherent cells was removed. The cells, attached to the plastic on the bottom of the culture flask, were gently washed with DPBS, the medium was completely replaced. Subsequent medium replacements were carried out every 2-3 days; the cultures were examined using phase-contrast microscopy.

As the culture was cultured and the subconfluent state reached, the cells were trypsinized with a solution of Trypsin-EDTA ("Gibco") and passaged 1: 2. For the experiment, we used a culture of passage 3, 1 million cells were placed in a tooth socket.

A frequency-stabilized Helium-Neon laser with two orthogonal optical modes was used to transfer the quantum genetic information, which read the genetic information from the rudiment of the human tooth. Such information was spontaneously transformed into modulated broadband electromagnetic radiation (MBER) carrying the same information, initially recorded on polarization modulation (spin states) of the probing photons in the mode of returning the laser beam back to its resonator. The theory of this process was published by us in [4]. Particularly, MBER was synthesized from the surgically removed rudiment of a human molar. Multipotent mesenchymal stromal cells (MMSCs) were extracted from human adipose tissue. These MMSCs were irradiated with the synthesized MBER and were grown to the required concentration.

The dog's teeth were removed behind the fangs on the left and right side of the jaw. A week later, multipotent mesenchymal stromal cells, pre-treated with MBER of the human tooth rudiment, were implanted in the place of the removed right tooth. The control area from the removed left tooth did not receive any cell implantation.

The test dog was handled in accordance with the accepted standards in scientific research in Russia.

This procedure resulted in regeneration of the new teeth over 9 months. In the control area of the left tooth, there was no regeneration:

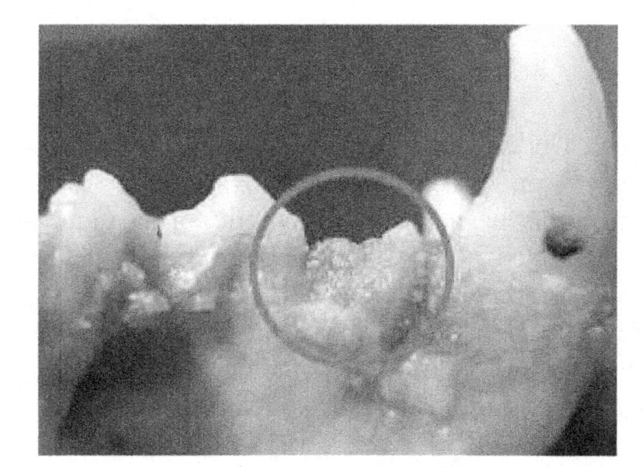

Figure 1A. Control: The photograph shows the left upper jaw of the test dog where the tooth was removed, no multipotent mesenchymal stem cells (MMSCs) were implanted, and there was no regeneration of tooth.

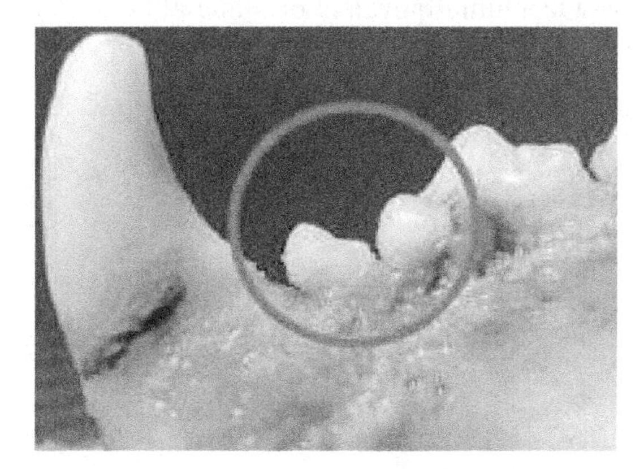

Figure 1B. Test: The photograph shows the right upper jaw of the test dog where the tooth was removed, multipotent mesenchymal stem cells (MMSCs) were implanted, and there were regenerations of teeth after 90 days.

Conclusion

In this paper, we reported our preliminary study demonstrating tooth regeneration in the test dog using human genetic information. Regeneration was carried out by a special laser technology based on an expanded understanding of the principles of genetic coding. One of such principles predicts the existence of quantum equivalents of working genes. The work reported here will be repeated and how human genetic information interacts with canine genetic information will be explored.

References

1. Gariaev P.P., 2015, Another Understanding of the Model of Genetic Code. Theoretical. Analysis. Open Journal of Genetics, v.5, pp. 92-109.
2. Gariaev P.P., 2018, Leonova-Gariaeva E.A., The Syhomy of the Genetic Code Is the Path to the Real Speech Characteristics of the Encoded Proteins. Open Journal of Genetics, v.8, №2.
3. Gariaev P.P., Poltavtseva R.A., Leonova-Gariaeva E.A., Voloshin L.L., Dobradin A., 2017, Practical Application of Linguistic Wave Genetics (LWG) Principle in creating Quantum Information Matrices (QIM) used for Programming Plain Liquids into Medically Active Liquids, called Quantum Information Matrix Programmed Liquids (QIMPL). Clinical Epigenetics. 3:22. doi: 10.21767/2472-1158.100056.
4. Prangishvili I.V., Gariaev P.P., Tertyshny G.G., Maksimenko V.V., Mologin A.V., Leonova E.A., Muldashev E.R., 2000, Spectroscopy of Radio-Wave Radiations of Localized photons, Sensors and Systems, No. 9 (18), pp.2-13 (In Russian).
5. Gariaev, P. P., Vladychenskaya, I. P. & Leonova-Gariaeva, E. A., PCR Amplification of Phantom DNA Recorded as Potential Quantum Equivalent of Material DNA. DNA Decipher Journal, v.6, issue 1 (March 2016), pp. 1-11.

Article

Time, Life & the Emotive Source

Stephen P. Smith*

Abstract

A Panpsychism, or neo-vitalism, is presented having to do with the penetration of time in living organism. Time is described having bifurcated or polarized into two windows: one that looks forward in time and follows a chain of determinism, and one that looks backward in time to frequencies and past habits. The emotive source is described as a singularity, the timeless middle-term holding the two windows together. This view is related to the laws of physics, the second law of thermodynamics, genetics and epigenetic switches. Warm-body quantum mechanics is implicated broadly, and in particular with the creation of adaptive mutations that are coxed by epigenetic cues.

Keywords: Time, life, emotive force, panpsychism, neo-vitalism, living organism, determinism, bifurcation, thermodynamics, genetics, epigenetics, quantum mechanics, creation.

1. Introduction

The visionary and counter-culture proponent[1], Terence McKenna (1946-2000), defined life as something that time got into.[2] Saw the legs off a chair, and throw the pieces on the ground, and nothing will happen, he noted. Come back the next day, and the pieces are still there where you left them, unchanged. Do that with something living, he noted, and it will bleed and die as time unfolds. Cutting body parts off of something living, he concluded, is to interrupt time and its connection to life.

In this paper, McKenna's hypophysis is taken serious, but without any intended cruelty of animals. Surprisingly so, his view is found very compelling, that time has gotten into life. The view sees time bifurcated in life into two windows. From life emerged mind and the view see time bifurcated in our psychology. The two-sided time implies a timeless middle-term, hinting of a panpsychism or the emotive source.

The philosophical proposal of panpsychism, as well as scientific accounts of consciousness based on quantum mechanics, are becoming more acceptable today as serious endeavors. To these we can add the emerging field of quantum biology. Regarding biology, its reasonable to assume that panpsychism becomes a neo-vitalism if there is any truth to panpsychism. To proceed, this paper reissues the second law of thermodynamics in Section 2 in terms that are friendly to neo-vitalism, without diminishing the law-like drive that increases entropy with time

Correspondence: Stephen P. Smith, Ph.D., Independent Researcher. E-mail: hucklebird@aol.com Note: This article was first published in Journal of Consciousness Exploration & Research 9(8): pp. 707-721 (2018).

1 This paper makes no endorsement of McKenna's advocacy of psychedelics.
2 By comparison, Immanuel Kant viewed time as an internal intuition.

passage. Genetics is described in Section 3, but in summary form. An account of time-impacted life based only on genetics, while missing epigenetics, is stilted at best. Therefore, Section 4 describes epigenetics, but in summary form. To relate neo-vitalism to what known about physics, including the 2nd law, it is necessary to reissue all these laws in Section 5. An account of how time possibly leaves its mark on epigenetic switching is presented in Section 6, including a universal grammar that engages proto-emotion and hinting of a necessary warm-body quantum mechanics. This leads to the possibility of some testable science, and a discussion of such possibilities are presented in Section 7.

2. Second Law of Thermodynamics

The second law of thermodynamics describes the fall of energy from an ordered state (low entropy) to a disordered state (high entropy), in a universe where energy is otherwise conserved. The 2nd law is irreversible, in that energy has never been observed to flow in the opposite direction, from a disordered state to an ordered state that's more that just a minor occurrence. A good example is a ball that is perfectly balanced on top of a hill (Figure 1). The balanced state represents a perfect symmetry, where the path to be taken in the down-hill role is still not determined. This symmetry represents an impartiality in the future possibilities, but its highly unstable. The slightest wind will tip the balance and break the symmetry, causing the ball to start rolling in some particular direction. Initially, the ball has high potential energy representing the a-prior order. As the ball starts rolling, the potential energy deceases and is transformed into some waste heat and kinetic energy when the ball accelerates. Eventually that ball will come to rest in one of four valleys, and at that point all of the incremental change in potential energy is transformed into waste heat. The one-way flow has never been observed to go in reverse, where the ball absorbs waste heat from the surroundings and rolls up hill, and then finds itself perfectly balance at the apex of the hill.

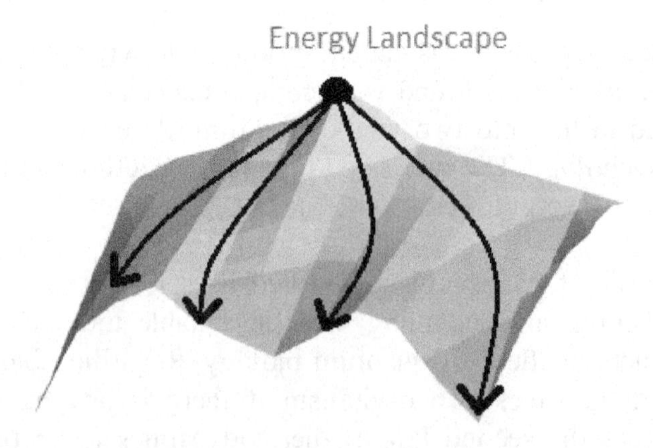

Figure 1. Energy landscaped showing black ball balanced on top of a hill, ever-ready to be tipped and to roll into one of four possible gullies before coming to rest in one of the respective valleys.

Statistical mechanics can represent the 2nd law as the behavior of free bodies inside a closed ensemble.[3] The free bodies can float around unrestricted, in random directions, bouncing off each other like Newtonian billiard balls. The migrations are restricted by transitions that are determined by probability. If all the free bodies are located in one corner of the ensemble initially, neatly stacked side by side, then that would represent a highly ordered state. With time passage the ordered state would decay and fall into a disordered state where the free bodies are smeared out in any such direction, almost uniformly over the space of the ensemble but deeply haphazardly. The 2nd law dictates that energy transitions from low entropy to high entropy, and never in reverse, and if you believe statistical mechanics this one-way flow is a matter of probability.

When the probabilistic version of the 2nd law is interpreted as a universal law, however, what is discovered are deep contradictions having to do with the low entropy of the initial conditions (Price 1996, Chapter 2). This made Boltzmann and others rethink the probabilistic law as a universal. If thermodynamic equilibrium is the normal state of affairs when the universe began Boltzmann figured, then statistical mechanic paradoxically predicts that entropy was much higher at some time in the distant past than today and coming with high probability, a direct contradiction of the 2nd law (Price, page 30). Or we can assume that the initial conditions, now given as the birth of the universe, came with low entropy (Albert 2000, Chapter 4). Where did that initial high order come from?

Smith (2008) described the 2nd law as two-sided, and holding the fatal equivocation of the meaning of "representation" given by statistical mechanics and the meaning of "recognition" given by a space that dissipates waste heat. William James Sidis (1925) made similar observations, and predicted a new principle that works in reverse of the 2nd law and existed in some pockets of the universe, and related this new principle to life and life's teleology[4].

Given that Kauffman (2008) describes a "ceaseless creativity" emerging at the criticality separating order and disorder, the 2nd law described by statistical mechanics can only be a one-sided interface showing the fall into disorder. The other side of the demarcation is where disorder is found uniting into order again, forming larger wholes, but agreeing with Sidis (1925) this must necessary happen in reverse-time to maintain consistency with both sides. Moreover, Sidis found the 2nd law to be a psychological law that's necessary to keep track of apparent forward causation. David Hume believed that our understanding of causation came to us by regularities discovered with past experiences rather than by reason.[5] Time is found polarizing into two psychological windows, then. One side looks forward deductively and recognizes deterministic chains as Sidis found, but the other side looks backward to find itself in inductive habits, generalities and past frequencies as Hume believed. The two windows of time are necessarily unified by a middle-term that's undeclared by law, a middle-term that's necessarily timeless[6] and

---

3 This formulation is due to Ludwig Boltzmann, coming from his address to a meeting of the *Imperial Academy of Science*, 1886.

4 In a similar vein, Price (1996) also considers causal asymmetries implied by the 2nd law and postulated a possible backward causation or advanced action that involves quantum mechanics.

5 *An Enquiry Concerning Human Understanding.*

6 In the sense that photos are timeless (Russell 2003). Photons act as messenger particles that communicate the electromagnetic force, and are also massless and spaceless. Gluons and postulated gravitons are also massless and also act as messenger particles for the strong force and gravity,

represents a deep singularity. The middle-term is also implicated as the emotive source[7]; i.e., if Alfred North Whitehead is correct in his belief that causality can be directly prehended by experiential occasions, however vague a feeling but an emotion laden feeling nevertheless.[8]

The 2nd law describes the behavior of mindless free-bodies that risk the heat death, and Sidis's psychological law translates into the emotional warning: remain mindful or bad things will happen. It's the kind of advice that a father will give to his son to stay on the straight and narrow path. It's the watchful eye of the shepherd that looks after the flock and plans ahead. Hume's version of the same law, but on the receiving side, is more feminine: we are all in this together; and harmony is preferable to strife.

3. Genetics and the Fitness Landscape

Wright (1932) was the first to describe the fitness landscape for a population where its average fitness is given by a position on a topology that's dependent on gene frequency. Natural selection and genetic drift can both influence gene frequency changes that can occur in a population over time, from one point on the topology to the next, presumably climbing higher on the landscape. Figure 2 provides an illustration of a simple fitness landscape that applies on the population level, but there are other varieties of fitness landscapes that have been studied (e.g., Kauffman 1993).

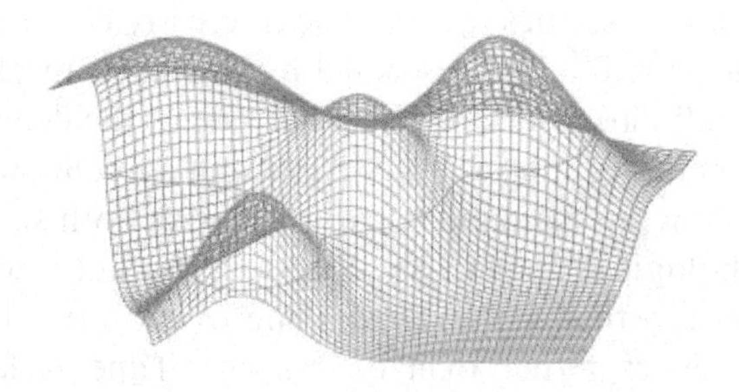

Figure 2. Hypothetical fitness landscape given by the height dimension and overlaying a two-dimensional surface that depicts gene frequency at two independent loci.

Independent of how variation offers itself to natural selection, the fitness landscape drives evolution. It represents a genetic determinism that points to a possible future state of a population that is more adapted. However, the fitness landscape is just one possible driver of evolution

respectively.

7 Rather than describing this as a panpsychism, or a panexperientalism, preference is given to proto-emotionality as the fundamental because this is arguable less anthropomorphic and yet it carries the essential meaning that preferred directions are sought non-passively.

8 *Process and Reality*, Chapter VII.

among genetic drift, genetic recombination, mutation and the survival instinct that's innate in life. It's a gross oversimplification to imply these processes are blind and indifferent, particularly if evolution is to now be connected to the mysterious 2nd law of thermodynamics (e.g., Wicken 1987; Brooks and Wiley 1988; Chaisson 2001), particularly if life's adaptability is found relating to quantum mechanics and the collapse of the quantum wave function (Goswami 2008; McFadden 2000), and particular if mutations are found not random but life-directed and even adaptive (Cairns, Overbaugh and Miller 1988; Martincorena and Luscombe 2013). Mutations that are accidental, or haphazard, tend to be detrimental to such an extent that natural selection is unable to remove them in numbers that can improve the adaptation of living organisms (Stanford 2008, Chapter 4). It is necessary for cells to utilize error correcting capabilities when mutations, or copying errors, occur in DNA (Radman and Wagner 1988).

The fitness landscape joints the energy landscape (Figure 1) in that both imply a direction in time. Genes found in the zygote somehow predict heritable traits that may be selected in the adult by natural selection. In that evolution of novelty can emerge unexpectedly from Darwinian pre-adaptions (Kauffman 2008), the fitness landscape must be able to coax the novelty into existence as if the landscape carried a remarkable foresight. Nevertheless, the fitness landscape is a tool of the one-sided rational mind that sees a chain of causation in a world well described by Newtonian billiard balls that bounce off of each other.

4. Epigenetics and the Waddington Landscape

Epigenetics controls developmental biology by way of switches that are actually put on the DNA. The switches turn genes off with methylation (Siegfreid and Cedar 1997), or they slow or accelerate the gene with a connection using histones by acetalyzation (Eberharter and Becker 2002). In other words, the epigenetic switches act as frequency modulators on the DNA by determining the frequency profile of protein production.

The genetic material, or DNA, was never well described as a blue print or program. Epigenetics reveals that genetic determinism is better described by the metaphor provided by a Fourier analysis of a time series that depicts gene function as an action in time, where the time series is described as a linear combination of a set of basis vectors (now the original DNA backbone), and where the linear coefficients are spectral frequencies attached to each gene by epigenetics.

After fertilization the zygote divides and becomes a blastocyst, differentiated only into an inner cell mass (ICM) and the outer trophectoderm. The ICM and the trophectoderm will differentiate further into the embryo and placenta, respectively. When ICM cells go through differentiation they specialize into tissue types, liver, brain, blood, hair, etc. While each cell in the body has the same DNA, they have a different combination of epigenetic switches. In the process of cell division, these switches are passed on to the daughter cells, i.e., the switches are more the less permanent. However, the daughter cells can have new switches added if the daughter cells differentiate further from the parent, in a process that is not reversible[9]; i.e., the new epigenetic

9 There are notable exceptions. When animals are cloned, or when stem cells are generated experimental

switches are added during some of the steps of embryonic development where changes among cell types occur. The differentiation of cells during early development, starting with the zygote, resembles the irreversible fall into disorder having to do with the 2nd law of thermodynamics (Figure 1), a pattern of future possibilities provided by the Waddington landscape depicted in Figure 3. This characterization of irreversible differentiation as a ball rolling down an epigenetic landscape is due to Waddington (1957). Ferrell (2012) relates the dynamics carried by the Waddington landscape with an attractor, an observation that is not lost in Section 6.

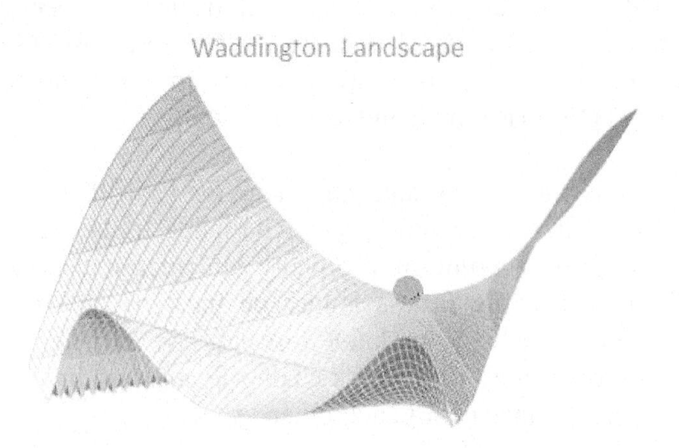

Figure 3. One cell (dark dot) from a blastocyst set to roll down the Waddington landscape and undergo irreversible differentiation following one of three paths that are provided by the two forks shown here. When the cell undergoes division it then assumes the identity of one of the daughter cells artificially, thus permitting a logical passage down the entire landscape.

Cells in the blastocyst are possibly in some still unexplored quantum entangled state, particularly if differentiation is recognized as partially a collective process but admitting to possible degrees of entanglement. This implies that daughter cells remain quantum entangled after cell division. With this interpretation, the blastocyst represents a quantum supper-position of possible realizations, differentiation being the collapse of respective wave functions when epigenetic switches are added to DNA (e.g., Asano et al., 2017). Jorgensen (2011) also implies that epigenetics involves quantum mechanics, even suggesting a backward in time flow of information from the environment.

Epigenetics is more complicated than presented in this summary, and a broad overview of the emerging subject is presented by Carey (2012) that also includes a description of the following side notes:

1. Some epigenetic switches can also be passed through sexual reproduction (e.g., Migicovsky and Kovalchuk 2013). A parent can acquire an epigenetic change that makes its way to the germ line, then past that change to a child that inherits the same characteristics in a Lamarckian sense.
2. Sperm and egg cells can also be imprinted differentially by epigenetic switches, in a way

---

from non-stem cells, the epigenetic switches are removed somewhat where experimental methods are discovered how to reverse epigenetic changes.

that's essential for development.

3. There is epigenetic regulation having to due with the deactivation of one X-chromosome, a necessary regulation for females that carry two X chromosomes.
4. The non-protein coding DNA, which is 98% of the DNA, transcribes into RNA that acts in epigenetic switching. These processes are not well understood.

5. Two-sided Laws and Quantum Mechanics

Einstein's special and general relativity are symmetric in time. The action principles of unified field theory look identical under CPT inversion[10], which includes time. In summary, all of the laws of physics are found symmetric in time as Sidis (1925) noted, apart from the asymmetrical 2nd law that is arguably two-sided. That is, the laws that come as action principles look the same in reverse time as they do in forward time, and they even hint of a possible teleology (Helrich 2007). Teleology can be recovered from the two-sided 2nd law as hypothesized in Section 2.

Something must also be said about symmetry breaking, because laws that look the same in both time directions, that look the same in their own specified symmetry so polarized to provide two views, say very little about how the laws themselves came into being by a process of symmetry breaking. In the grand unification epoch, the universe is thought to have expanded and cooled from a big bang (Chaisson 2001, pp 242-247). During this expansion, the four forces of nature emerged in the order gravity, the strong force, weak force, electromagnetism. This process involved spontaneous symmetric breaking, leaving the laws behind but coming with selected constants that modulate and calibrate the action of the laws. In the wake of these breaks we find: the speed of light, the cosmological constant for a flat universe, the relative strengths of the four forces of nature, the masses and coupling constants, Planck's constant. These are the affinities that nature selected. Most of these are thought fine-tuned for life (Barrow and Tipler 1986). The question comes, were these selections the result of random occurrences when symmetries broke, or were they innate preferences selected by the ground of being? The Weak Anthropic Principle tries to enforce the belief that it is all accidental, because we would not be here to ask questions if the selections were different, we are just very lucky.

The symmetrical laws themselves look to be taking part of a greater symmetry breaking once a boarder view is taken on how they act. The broader view changes with law-restrained action, but something remains that's unchanged called the law. First understand, that from the point of view of agency that carries information, symmetry is when everything looks the same from all points of view. After symmetry breaks things stop looking the same as natures arrives at an affinity or preference, and this implies there needs to be some kind of polarity before symmetry breaking that holds the two sides together in an indistinguishable state. Symmetry breaking leads to a state of natural discernment, where one side becomes visible while the other disappears. In this view, that part of reality that became invisible does not mean necessarily that the unselected part of reality stops existing. Because the middle-term is beyond law and cannot be excluded from reason, it only means that the unselected part of reality becomes part of a shadow. For example,

10 CPT is an acronym for Charge, Parity and Time. CPT inversion changes a particle into its anti-particle, changes an image into its mirror reflection, and changes forward time into reverse time.

the existence of dark energy and dark matter may be the left-over shadow that followed in the wake of symmetry breaking occurring in the early universe.

The broader view of symmetry breaking may very well be synonymous with the collapse of the quantum wave function, in these cases the collapse of the universe's own wave function by the action of gravity. This view is a restatement of John Wheeler's (1990) particpatory universe, where the constants of physics were selected by the ground of being because something was preferred to nothing; i.e., preferred by proto-emotion that connects time over a very wide duration. The collapse of the quantum wave function, more localized in brain tissues and involving microtubules in cells, has already been implicated as something fundamental to consciousness (Hameroff and Penrose 2014). Strapp (2007) also implicates the quantum wave function and its collapse to the phenomenon of consciousness, but Strapp points to the quantum Zeno effect within synapses rather than microtubules, and Strapp further relates his theory to Whitehead's process philosophy. Perhaps unsurprisingly now, the collapse of the quantum wave function may be the real driver behind irreversibility and the 2nd law (Albert 2000, Chapter 7), implying that order and disorder relate directly to quantum coherence and decoherence.

If our experience of free will is real, then it must be possible to transcend the apparent flow of forward causation, to reach back and set initial conditions by a reverse causation that is modulated by our emotions. Otherwise, Benjamin Libet's (1985) timing experiment will spell the end of our imagined freewill. Freewill is saved by quantum mechanics again, because its quantum mechanics that permits a reverse causation over a time duration (Wolf 1998).

In summary, the laws of physics are found two-sided, and the middle-term that holds the sides together is undeclared by law and is possibly the source of proto-emotion. Therefore, the historical and premature rejection of vitalism is now recognized as the fallacy of excluded middle. The possibility of reverse-mode causation necessarily implicates quantum mechanics, and the collapse of the wave function. For the same reasons, quantum mechanics is implicated in gene action and epigenetics, in all probability.

6. When Time Finds Itself

Up to this point, a mechanism has not been described how proto-emotion that's completely confounded with time is found relating to genetics and epigenetics. The proposed hypothesis is that the mechanism is part of a universal grammar given that time presents two windows. Only with the universal grammar better described can evidence be gathered to support this hypothesis.

The backward in time window relates to a signal sensitivity given by past frequencies that modulate the effect of DNA. This modulation is precisely what is found with the epigenetic switches provided by methylzation and acetalyzation. Epigenetics is not limited by the DNA modulation, however, because something epigenetic may be on top of DNA modulation, and something may be on top of that.[11] At any rate, the backward in time window is closely matches

11 Lipton (2005, Chapter 3) describes proteins that are embedded in the cell membrane, and act as information gates that relate to the outside environment and also impact on epigenetic signals inside

with the *frequency domain* that characterizes time series analysis. The other type of time series analysis is called the *time domain* which is characterize by a forward progression such as the Markov chain.

The time domain or the forward in time window has to do with the forward flux of cause and effect given by the genetic determinism that presents itself: be it a population that's navigating the fitness landscape; or cells undergoing division and embryonic differentiation by navigating the Waddington landscape; or it's a cell involved with its mature metabolic function thereby following an energy landscape.

The two time windows will permit negative feedback in forward time, where the chain of determinism may reach back and reset epigenetic switches. This is classically defined regulation that still represents something habitual. However, this is not the emotive balancing act described next that will actually carry a reverse-time connection.

The more speculative part of the grammar has to do with the creation of new functions. This is described as the great centering of proto-emotion to arrive at the critical point separating order and chaos.[12] It must be that the backward and forward windows come together as time finds itself in a process of targeted synchronisation, the two widows oscillate. This brings a tension leading to an emotive climax where something new erupts on time's manifold and comes into being so targeted: a new mutation or an alteration in an epigenetic switch. This coming into being involves the collapse of the quantum wave function so hypothesized, with a time duration and quantum non-locality that is significant enough to likely enhance the utility of the targeted change that was sought by proto-emotion.[13]

Strict or quasi genetic and epigenetic determinism[14] is inadequate during periods of crisis by definition, leading to exhaustion, stress, and even proto-emotional surrender. The response that comes in the wake of the eventual release is hypothesized as part of the universal grammar, which is the aforementioned coming together and synchronisation of time's windows, and this grammar is followed by the cell and by human psychology[15]. There is known genetic regulation that can kick into gear during periods of stress, inducing adaptive mutation (e.g., Hopkins et al., 2013; Chen, Lowenfeld and Cullis 2009; Tadderi et al., 1997). But is this the result of relaxing

---

the cell. Moreover, Pert (1997, Chapter 9) postulates that peptides circulating in the body communicate emotions.

12 What Vattay, Kauffman and Niiranen (2012) describes as the "Poised Realm".

13 Specifics are lacking, but particulars of warm-body quantum mechanics are better provided by future investigations. Unknown are where on DNA, RNA or proteins are regions of quantum supper-position? Is a quantum collapse limited to biochemicals inside one cell? Or are there any non-locality and entanglement effects among cells, among organisms?

14 A strict-determinism is a one-sided set of entailments that enforce the principle of excluded middle as Kauffman (2014) describes, whereas a quasi-determinism permits an emotive middle-term that has achieved a high level of fidelity in a learned route. Given that all physical laws are context dependent and are two-sided with undeclared middle-terms, this distinction is probably moot. Nevertheless, a learned route that involved something emergent (beyond known laws) is probably better characterized as a quasi-determinism.

15 Benjamin Franklin is quoted, "We must all hang together, or assuredly we shall all hang separately."

the error correcting capabilities of the cell as Keller (2000, page 34) implies? Where genetic stability and mutability are otherwise held in a delicate balance (e.g., Chen et al., 2010) because they are complementary epigenetic effects that modulate the same genes? And what is "stress," if not a non-preferred state from the point of view of proto-emotion? The mere fact that these serious questions can be raised in the face of the aforementioned references in primal facial evidence that supports the contention that mutability and genetic stability is the work of proto-emotional centering and synchronisation leading to switching. If the delicate balance is broken because of an innate one-sidedness that has developed, then the emotive centering may fail.

Note that mothering of baby rats creates a mood disposition in adult rats, all related to epigenetic methylation found in the brain that impacts on how adult rats relate to stress (Weaver et al., 2004). Rats can either be calm and tranquil, or easily agitated, when reacting to a stressful situation. Epigenetics is again found cultivating the hypothesized proto-emotion, but now to determine an emotional response in adult rats. The recurring balancing act implies that proto-emotion is necessarily two-sided, corresponding to time's two windows. This regulation reaches into human emotions. Mindfulness meditation, that will necessarily bring with it emotive centering, can improve health by impacting on epigenetic switching (Kaliman et al., 2014).

Not that the great centering is limited to periods of stress and new creations, rather once a route is explored the coming-together that led the way can give itself over to genetic and epigenetic quasi-determinism and become a learned route as part of the Waddington landscape. Gould (1977) describes how ontogeny recapitulates phylogeny, perhaps only as an approximation, but an approximation made possible because the proto-emotive centering that occurred in biological evolution also gave its self to the Waddington landscape to be expressed as a quasi-determinism. Mitosis describes chromosomes that double in number during interphase, and line up on the equatorial plane during metaphase, that is, they center. The cell divides as chromosomes are pulled by spindle fibers (aggregates of microtubules), this being the eruption on time's manifold coming with the emotive climax. Meiosis is more complicated, but also involves centering on the equatorial plane. Arguably, mitosis and meiosis have become part of the genetic and epigenetic quasi-determinism, taking with them emotive centering.

Lastly, there is an issue that relates to how epigenetic layers may stack one on top of the other. Perhaps the layers are all related by this syntax, an alternating pattern: frequency modulation leading to a localized determinism (more strict than quasi), on top is more frequency modulation and an additional implied layer of determinism, etc. This layering is open to experimental investigation and discovery, but somehow they must all be connected by warm-body quantum mechanics. Furthermore, the layering cannot be replaced by an overlaying strict-form determinism that ignores the emotive middle-terms that represent jumps beyond one-sided entailments, because those bridges cannot be "pre-stated" as Kauffman warns (2014); at best, the layers can only be united by a quasi-determinism. Therefore, finding such a layering experimentally that is alternating (from frequency to determinism) hints of the deep singularity connecting to the timeless.[16] A quantum system poised on the edge of order and chaos (and

16 Critics may even concede such a layering, if only because frequency switches followed by an implied determinism is all that has ever been discovered. However, it is possibly difficult to experimentally differentiate between strict and quasi determinism.

coincidentally between coherence and decoherence) can support warm-body coherence of quantum waves over an extended duration, by synchronizing an incoming wave-form to system frequencies that swing between regularity and chaos making a system resonance (Vattay, Kauffman and Nirranen 2012). Hameroff (2014) sees the vibrations found in microtubule supporting warm-body coherence, and permitting the reality of conscious experience. Vibrations are needed in a "goldilocks zone" to maintain coherence, and re-coherence, a goal that is in reach of biological evolution (Lloyd et al., 2011). To the extent that the eventual fall into decoherence represents an implied determinism, we find that frequency modulation does support a variety of determinism, even a sequence of restrained Zeno effects representing transitions (McFadden and Al-Khalili, 2016, pages 292-298). The entire system must necessarily be composed of multiple layers, however, assuming that warm-body coherence is more widespread in the body. This stacking of layers is agreeable with Kauffman's (2014) alternative quantum mechanics, where new "possibles" (i.e., new states of coherence) present themselves in the wake of "actuals" that form (when decoherence arrives).

7. Discussion

Section 6 can be seen as an alternative interpretation of quantum mechanics having to do with the collapse of the quantum wave function as the synchronisation of time's widows, when time looks back to find itself as an emotional interplay. However, the main fucus should remain on finding evidence, supporting the hypothesis or otherwise refuting it, and looking to alternative views of quantum mechanism may be less productive given all the varieties already described by Herbert (1985). This is not to say that what is proposed is radically different or incompatible with Goswami, Reed and Goswami's (1993) monistic idealism, or even with Kauffman's (2014) understanding that admits to a possible preference-laden panpsychism. As with Kauffman's proposed theory that carries the very significant triad[17], the hypothesis described in this paper is open to some testable science.

Warm-body quantum phenomena have already been found in biology (McFadden and Al-Khalili 2014); i.e., beyond the enzymatic reactions that utilize electron and proton tunneling. Photosynthesis is found involving coherent waves (rather than particles) that are able to efficiently reach reaction centers in green bacteria (Engle et al., 2007). The migratory pattern in birds may relate to quantum entanglement (Ritz et al., 2004). The sense of smell is possibly a quantum assisted phenomenon (Brookes et al, 2007). Microtubules have quantum properties enough to strongly implicate them in the orchestrated objective reduction model of consciousness (Hameroff 2014; Hameroff and Penrose 1996). The fact that quantum biology has become an entrenched science indicates some support for the ideas in Section 6. So there is general support for Section 6 given that quantum biology is real, and given references already sited in Section 6. Specifics are missing that relate to the proposed quantum-level action, and those are a matter of future investigations. Nevertheless, if consciousness is now so strongly implicated with quantum mechanics in the brain, implying even a preference-driven panpsychism, one has to wonder how such a system could evolve without a parallel neo-vitalism

17 Actuals, Possibles and Mind represent a very important advance beyond scientific monism and Cartesian dualism.

of a kind hinted at in Section 6? Seems unlikely!

The now vindicated Terence McKenna believed that evolution was taking us to an endpoint. His "transcendental object at the end of time" was acting as a strange attractor and pulling us into the future and our end of days. He was possibly wrong about the attractor being stuck at the end of time, when proto-emotion is possibly timeless and no less a strange attractor. What's nice about strange attractors is that they leave evidence behind in the substrate of evolution given as reflections, on all levels being a fractal pattern. Therefore, we might look to the evidence coming as reflections of the timeless pull on time's two windows, to support the view of Section 6. Remarkable so, we do find such evidence. The philosophy of Taoism is remarkable on how it relates to our psychology, and how it relates to the more deductive Yang and the more inductive Yin,[18] all agreeing more or less with Section 6 as a universal grammar. This reflection is found impacting depth psychology of a kind promoted by Peterson (1999). Not merely is the reflection found in psychology, but also in the way our asymmetrical brain is wired together in a left and right hemisphere that specialized differently to look forward and backward in time, respectively. The brain's asymmetrical hemispheres are well described in McGilchrist (2009).

Religious traditions based on love[19] relate well to the timeless pull of a proto-emotionality, that comes with a universal grammar turned golden rule, and comes with a relational view of the world that agrees more or less with Whitehead's process philosophy.

A closing question: what came first, genetics or epigenetics? The possibility is that they co-evolved together, in a grand centering near the criticality where time was able to find itself through a process of synchronization. If the middle-term is the pristine source of all that is emotive, then the centering of proto-emotion is part of a universal grammar and far from a cheap variety of post-modernism. This implies that evolution carried its own direction and was highly non-passive, far from indifferent and blind as implied by Dawkins (1996).

References

Albert, D.Z., 2000, *Time and Chance*, Harvard University Press.

Asano, M., I. Basieva, A. Khrennikov and I. Yamato, 2017, A Model of Differentiation in Quantum Bioinformatics, *Progress in Biophysics and Molecular Biology*, 130 (Part A): 88-98.

Barrow, J.D., and F.J. Tipler, 1986, *The Anthropic Cosmological Principle*, Oxford University Press.

Brookes, J.C., F. Hartoutsiou, A.P. Horsfield and A.M. Stoneham, 2007, Could Humans Recognize Odor by Photon Assisted Tunneling?, *Physical Review Letters*, 98 (3): 038101(1-4).

Brooks, D.R., and E.O. Wiley, 1988, *Evolution as Entropy: Toward a Unified Theory of Biology*, 2nd Edition, The University of Chicago Press.

Cairns, J., J. Overbaugh and S. Miller, 1988, The Origin of Mutants, *Nature*, 335: 142-145.

---

18 The connection of deduction to the masculine, and of induction to the feminine, is better described in Smith (2010). Smith also described how the two tendencies may oscillate as a subjective experience.

19 St Augustine describes a relational view of the Trinity from the idea that God is Love (from Book 9 in *On the Trinity*).

Carey, N., 2012, *The Epigenetics Revolution*, Columbia University Press.

Chaisson, E.J., 2001, *Cosmic Evolution: The Rise of Complexity in Nature*, Harvard University Press.

Chen, Y., R. Lowenfeld and C.A. Cullis, 2009, An Environmentally Induced Adaptive (?) Insertion Event in Flax, *International Journal of Genetics and Molecular Biology*, 1 (3): 38-47.

Chen, F., W.-Q. Liu, A. Eisenstark, R.N. Johnston, G.-R. Liu and S.-L. Liu, 2010, Multiple Genetic Switches Spontaneously Modulating Bacterial Mutability, *BMC Evolutionary Biology*, 10 (277): 1-11.

Dawkins, R., 1996, *The Blind Watchmaker: Why the Evidence of Evolution Reveals a Universe without Design*, W.W. Norton & Company.

Eberharter, A., and P.B. Becker, 2002, Histone Acetylation: A Switch between Repressive and Permissive Chromatin, *EMBO Reports*, 3 (3): 224-229.

Engel, G.S., T.R. Calhoun, E.L. Read, T.-K. Ahn, T. Mančal, Y.-C. Cheng, R.E. Blankenship and G.R. Fleming, 2007, Evidence for wavelike energy transfer through quantum coherence in photosynthetic systems, *Nature,* 446 (April 12): 782–786.

Ferrell, J.E, 2012, Bistability, Bifurcations, and Waddington's Epigenetic Landscape, *Current Biology*, 22 (11): R458-R466.

Goswami, A., R.E. Reed and M. Goswami 1993, *The Self-Aware Universe: How Consciousness Creates the Material World*, Penguin Putnam.

Goswami, A., 2008, *Creative Evolution: A Physicist's Resolution Between Darwinism and Intelligent Design*, Quest Books.

Gould, S.J., 1977, *Ontogeny and phylogeny*, The Belknap Press of Harvard University Press.

Hameroff, S.R., and R. Penrose, 1996, Orchestrated Reduction of Quantum Coherence in Brain Microtubules: A Model for Consciousness, *Mathematics and Computers in Simulation*, 40: 453-480.

Hameroff, S.R., 2014, *Consciousness, Microtubules, & 'Orch OR': A Space-time Odyssey, Journal of Consciousness Studies*, 21 (3-4): 126-153.

Helrich, C.S., 2007, Is There a Basis for Teleology in Physics?, *Zygon*, 42 (1): 97-

110.

Herbert, N.,1985, *Quantum Reality: Beyond the New Physics*, Anchor Books.

Hopkins, M.T., A.M. Khalid, P.-C. Chang, K.C. Vanderhoek, D. Lai, M.D. Doerr and S.J. Lolle, 2013, De Novo Genetic Variation Revealed in Somatic Sectors of Single Arabidopsis Plants, *F1000Research*, 2 (5): 1-16.

Jorgensen, R.A., 2011, Epigenetics: biology's quantum mechancis, *Frontiers in Plant Science*, 2 (Articel 10): 1-4.

Kaliman, P., M.J. Álvarez-López, M. Cosín-Tomás, M.A. Rosenkranz, A. Lutz and R.J. Davidson, 2014, Rapid Changes in Histone Deacetylases and Inflammatory Gene Expression in Expert Meditators, *Psychoneuroendocrinology*, 40: 96-107.

Kauffman, S.A., 1993, *The Origins of Order: Self-organization and Selection in Evolution*, New York, Oxford University Press.

Kauffman, S.A. , 2008, *Reinventing the Sacred: A New View of Science, Reason and Religion*, Basic Books.

Kauffman, S.A., 2014, Beyond the Stalemate: Conscious Mind-Body - Quantum Mechanics - Free Will -

Possible Panpsychism - Possible Interpretation of Quantum Enigma, arXiv achieved, paper # 1410.2127.

Keller, E.F., 2000, *The Century of the Gene*, Harvard University Press.

Libet, B., 1985, Unconscious Cerebral Initiative and the Role of Conscious Will in Voluntary Action, *The Behavioral and Brain Sciences*, 8: 529–566

Lipton, B.H., 2005, *The Biology of Belief: Unleashing the Power of Consciousness, Matter, and Miracles*, Mountain of Love/Elite Books.

Lloyd, S., M. Mohseni, A. Shabani, H. Rabitz, 2011, The Quantum Goldilocks Effect: On the Convergence of Timescales in Quantum Transport, arXiv achieved, paper # 1111.4982.

Martincorena, I., and N.M. Luscombe, 2013, Non-random Mutation: The Evolution of Targeted Hypermutation and Hypomutation, *BioEssays*, 35(2): 123-130.

McGilchrist, I., 2009, *The Master and his Emissary: The Divided Brain and the Making of the Western World*, Yale University Press.

McFadden, J., 2000, *Quantum Evolution: The New Science of Life*, W.W. Norton & Company.

McFadden, J., and J. Al-Khalili, 2014, *Life on the Edge: The Coming of Age of Quantum Biology*, Broadway Books.

Migicovsky, Z., and I. Kovalchuk, 2013, Changes to DNA Methylation and Homologous Recombination Frequency in the Progeny of Stressed Plants, *Biochemistry and Cell Biology*, 91 (1): 1-5.

Pert, C.B., 1997, *Molecules of Emotion*, Scribner.

Peterson, J.B., 1999, *Maps of Meaning: The Architecture of Belief*, Routledge.

Price, H., 1996, *Time's Arrow and Archimedes' Point: New Directions for the Physics of Time*, Oxford University Press.

Radman, M., and R. Wagner, 1988, The High Fidelity of DNA Duplication, *Scientific American*, 259 (2): 40-46.

Ritz, T., P. Thalau, J.B. Phillips, R. Wiltschko and W. Wiltschko, 2004, Resonance Effects Indicate a Radical-Pair Mechanism for Avian Magnetic Compass, *Nature*, 429 (May 3): 177-180.

Russell, P., 2003, *From Science to God*, New World Library.

Sidis, W.J., 1925, *The Animate and the Inanimate*, The Gorham Press.

Siegfreid, Z., and H. Cedar, 1997, DNA Methylation: A Molecular Lock, *Current Biology*, 7 (5): R305-R307.

Smith, S.P. 2008, *Trinity: the Scientific Basis of Vitalism and Transcendentalism*, iUniverse, Inc.

Smith, S.P., 2010, The Proclivities of Particularity and Generality, *Journal of Consciousness Exploration & Research*, 1 (4). 429-440.

Stanford, J.C., 2008, *Genetic Entropy & The Mystery of the Genome*, FMS Publications.

Strapp, H.P., 2007, *Mindful Universe: Quantum Mechanics and the Participating Observer*, Springer.

Tadderi, F., M. Vulić, M. Radman and I. Matić, 1997, Genetic Variability and Adaptation to Stress, in Environmental Stress, Adaptation and Evolution, editors R. Bijlsma and V. Loescheke, 271-290.

Vattay, G., S. Kauffman and S. Niiranen, 2012, Quantum Biology on the Edge of Quantum Chaos, arXiv achieved, paper# 1202.6433.

Waddington, C.H., 1957, *Strategy of the Genes*, George Allen & Unwin.

Weaver, I.C., N. Cervoni, F.A. Champagne, A.C. D'Alessio, S. Sharma, J.R. Seckl, S.I. Dymov, M. Szyf and M.J. Meaney, 2004, Epigenetic Programming by Maternal Behavior, *Nature* Neuroscience, 7 (8): 847–854.

Wheeler, J.A., 1990, Information, physics, quantum: The search for links, In *Complexity, Entropy, and the Physics of Information*, editor Wojciech H. Zurek, Addison-Wesley.

Wicken, J.S., 1987, *Evolution, Thermodynamics, and Information: Extending the Darwinian Program*, Oxford University Press.

Wolf, F.A., 1998, The Timing of Conscious Experience: A Causality-Violating, Two-Valued, Transactional Interpretation of Subjective Antedating and Spatial-Temporal Projections, *Journal of Scientific Exploration*, 12 (4): 511-542.

Wright, S., 1932, The Roles of Mutation, Inbreeding, Crossbreeding and Selection in Evolution, *Proceedings of the 6th International Congress of Genetics*, 1 (8): 355-366.

Article

Dark Valence Electrons, Dark Photons, Bio-photons & Carcinogens

Matti Pitkänen [1]

Abstract

The possible role of bio-photons in living matter is becoming gradually accepted by biologists and neuroscientists. Bio-photons serve as a diagnostic tool and it seems that their intensity increases in non-healthy organism. I have proposed that bio-photons emerge from what I call dark photons, which are ordinary photons but have non-standard value $h_{eff} = nh_0$ of Planck constant. In this article the consequences of the hypothesis that dark photons emerging from the transitions of dark valence electrons of any atom possessing lonely unpaired valence electron could give rise to part of bio-photons in they decays to ordinary photons. The hypothesis is developed by considering a TGD based model for a finding, which served as a starting point of the work of Popp: the irradiation of carcinogens with light at wavelength of 380 nm generates radiation with wavelength 218 nm so that the energy of the photon increases in the interaction. Also the findings of Veljkovic about the absorption spectrum of carcinogens have considerably helped in the development of the model. The outcome is a proposal for dark transitions explaining the findings of Popp and Veljkovic. The spectrum of dark photons also suggests a possible identification of metabolic energy quantum of .5 eV and of the Coulomb energy assignable to the cell membrane potential. The possible contribution to the spectrum of bio-photons is considered, and it is found that spectrum differs from a smooth spectrum since the ionization energies for dark valence electrons depending on the value of h_{eff} as $1/h_{eff}^2$ serve as accumulation points for the spectral lines. Also the possible connections with TGD based models of color vision and of music harmony are briefly discussed.

Keywords: Dark photon, biophoton, valance electron, carcinogen, TGD framework.

1 Introduction

The possible role of bio-photons in living matter is becoming gradually accepted by biologists and neuroscientists. It seems that the intensity of bio-photon emission increases in sick organisms and bio-photons are used as a diagnostic tool. Fritz Popp (see `http://tinyurl.com/y7assha7`) started his work with bio-photons with some observations about the interaction of UV light with carcinogens [2] (see `http://tinyurl.com/y76a9fo4`). Veljckovic (`http://tinyurl.com/yatedje8`) has also published results suggesting correlations between carcinogenity and the absorption spectrum of photons in UV (ultraviolet).

I have proposed that bio-photons emerge as ordinary photons from what I call dark photons, which differ from ordinary photons in that they have non-standard value $h_{eff} = nh_0$ of Planck constant[12, 13]. Also other particles - electrons, protons, ions,..., can be dark in this sense.

One of the mysteries of biology, which mere biochemistry cannot explain, is that living systems behave coherently in macroscopic scales. The TGD explanation for this is that dark particles forming Bose-Einstein condensates (BECs) and super-conducting phases at magnetic flux tubes of what I call magnetic body possess macroscopic quantum coherence due to the large value of h_{eff}. This quantum coherence would force the coherent behavior of living matter. I have already earlier developed rather concrete models for bio-photons [12, 13] on basis of this assumption.

In the sequel I will discuss bio-photons from a new perspective by starting from bio-photon emission as a signature of a morbid condition of organism. The hypothesis is that in sick organism dark photons

[1]Correspondence: Matti Pitkänen `http://tgdtheory.com/`. Address: Rinnekatu 2-4 A8, 03620, Karkkila, Finland. Email: matpitka6@gamail.com.

tend to transform to bio-photons in absence of metabolic feed increasing the value of h_{eff}. Hence BECs of dark photons and also of other dark particles decay and this leads to a loss of quantum coherence.

A further hypothesis is that at least a considerable part of bio-photons emerge in the transformations of dark photons emitted in the transitions of lonely dark valence electron of any atom able to have such. Since dark electron has a scaled up orbital radius, it sees the rest of atom as a unit charge and its spectrum is in good approximation hydrogen spectrum. Therefore the corresponding part of the spectrum of bio-photons would be universal in accordance with quantum criticality.

This picture allows to develop some ideas about quantum mechanisms behind cancer in TGD framework.

1.1 Some basic notions related to carcinogens

Before continuation it is good to clarify some basic notions. Toxins are poisonous substances created in metabolism. Carcinogens (`http://tinyurl.com/ybphtjqg`) are substances causing cancer, which often cause damage to DNA and induce mutations (mutagenicity).

Free radicals (see `http://tinyurl.com/y9bxoqjz`) provide a basic example about carcinogens. They have one un-paired valence electron and are therefore very reactive. The un-paired electron has a strong tendency to pair with an electron and steals it from some molecule. The molecule providing the electron is said to oxidize and free radical to act as oxidant. The outcome is a reaction cascade in which carcinogen receives electron but electron donor becomes highly reactive. Anti-oxidants stop the reaction cascade by getting oxidized to rather stable molecules (`http://tinyurl.com/omb7kc9` and `http://tinyurl.com/ydeloxcn`).

Benzo[a]pyrene (BAP) $C_{20}H_{12}$ (see `http://tinyurl.com/y8etnmwb`) is one example of carcinogen. It contains several carcinogenic rings and is formed as a product of incomplete burning and reacts with powerful oxidizers. As such BAP is not free radical but its derivatives $BAP^{\pm}$ obtained by one-electron reduction or oxidation are such (see `http://tinyurl.com/yb7am8tk`).

There are also carcinogens such as bentzene, which as such is not dangerous. What happens is that to the carbon at the ends of bentzene's double bond binds single oxygen atom and so called epoxy bond is formed. This molecule penetrates to the DNA chain and causes damage. Perhaps the fact that DNA nucleotide also contains aromatic 6-rings relates to this.

The emission of bio-photons (see `http://tinyurl.com/ol39rqx`) increases if carcinogens such as oxidants are present. The idea is that bio-photons could be relevant concerning the understanding of the problem. It has been proposed that bio-photons could be created when anti-oxidants interact with molecules generating triplet states (spin 1) which decay by photon emission. The photons generated in this manner would have discrete spectrum whereas bio-photons seem to have continuous and rather featureless spectrum. Therefore this model must be taken with caution.

It could be that the origin of bio-photons is not chemical. If so, carcinogens would not produce bio-photons in ordinary atomic or molecular transitions. They could be however induce generation of bio-photons indirectly. The understanding of bio-photons might help to understand the mechanisms between carcinogenic activity. I have discussed bio-photons from TGD view in [12, 13].

1.2 Some basic notions of TGD inspired quantum biology

In the sequel I try to develop a necessarily speculative picture about carcinogen action on basis of TGD based quantum about biology [10, 11]. The goal is to develop the general theory by developing a concrete model for a problem.

Magnetic flux tube and field body/magnetic body are basic notions of TGD implied by the modification of Maxwellian electrodynamics [10, 4, 9]. Actually a profound generalization of space-time concept is in question. Magnetic flux tubes are in well-defined sense building bricks of space-time - topological field quanta - and lead to the notion of field body/magnetic body as a magnetic field identity assignable to any physical system: in Maxwell's theory and ordinary field theory the fields of different systems superpose and one cannot say about magnetic field in given region of space-time that it would belong to some particular system. In TGD only the effects on test particle for induced fields associated with different space-time sheets with overlapping M^4 projections sum.

The hierarchy of Planck constants $h_{eff} = n \times h_0$, where h_0 is the minimum value of Planck constant, is second key notion. h_0 need not correspond to ordinary Planck constant h and both the observations of Randell Mills [18] and the model for color vision [23] suggest that one has $h = 6h_0$. The hierarchy of Planck constants labels a hierarchy of phases of ordinary matter behaving as dark matter.

Magnetic flux tubes would connect molecules, cells and even larger units, which would serve as nodes in (tensor-) networks [1][17]. Flux tubes would also serve as correlates for quantum entanglement and replace wormholes in ER-EPR correspondence proposed by Leonard Susskind and Juan Maldacena in 2014 (see `http://tinyurl.com/y7za98cn` and `http://tinyurl.com/ydckw5u7`). In biology and neuroscience these networks would be in a central role. For instance, in brain neuron nets would be associated with them and would serve as correlates for mental images [19, 24]. The dynamics of mental images would correspond to that for the flux tube networks.

1.3 The proposed model briefly

In the sequel the basic hypothesis will be that dark photons emerging from the transitions of dark valence electrons of any atom possessing lonely unpaired valence electron could give rise to part of bio-photons in they decays to ordinary photons. The hypothesis is developed by considering a TGD based model for a finding, which served as a starting point of the work of Popp (see `http://tinyurl.com/y76a9fo4`): the irradiation of carcinogens with light at wavelength of 380 nm generates radiation with wavelength 218 nm so that the energy of the photon increases in the interaction. Also the findings of Veljkovic about the absorption spectrum of carcinogens (`http://tinyurl.com/yatedje8`) have considerably helped in the development of the model.

The outcome is a proposal for dark transitions explaining the findings of Popp and Veljkovic. The spectrum of dark photons also suggests a possible identification of metabolic energy quantum of .5 eV and of the Coulomb energy assignable to the cell membrane potential. The possible contribution to the spectrum of bio-photons is considered, and it is found that spectrum differs from a smooth spectrum since the ionization energies for dark valence electrons depending on the value of h_{eff} as $1/h_{eff}^2$ serve as accumulation points for the spectral lines. Also the possible connections with TGD based models of color vision and of music harmony (see [23, 16, 26]) are briefly discussed.

2 About the modelling of the basic findings of Popp and Veljkovic

The popular article about starting point of Popp's research work (see `http://tinyurl.com/y76a9fo4`) tells that one can assign to carcinogens such as benzo[a]pyrene (polycyclic aromatic compound - a wave length $\lambda_i = 380$ nm. Carcinogen absorbs this wavelength and radiates photons with a shorter wavelength $\lambda_f = 218$ nm. In the following I try to understand what could happen in this process. I also consider the observations of Veljkovic [3] and their relationship to the findings of Popp.

2.1 General TGD picture

The zeroth order iterate for TGD interpretation of the action of free radicals would be following. Free radicals lead to the destruction of dark phases with non-standard value of h_{eff}. These phases include Bose-Einstein condensates of various kinds and super-conducting phases. The process leads to an emission of dark photons which transform to ordinary photons identified as bio-photons in the phase transition $h_{eff} \rightarrow h$. For instance, this happens as vegetable ageing and bio-photon emission is indeed used as a tool to determine the age of vegetable.

How the stealing of electrons by free radical electrons could induce the negative biological effects?

1. Quantum coherence is essential for what it is to be living matter. Bio-system is full of different kinds of Bose-Einstein condensates (BECs) and superconducting phases [6, 7]. Electronic superconductivity is one of the most important examples. There are also cyclotron BECs for proton Cooper pairs and biologically important bosonic ions or of the Cooper pairs of fermionic ions such as Ca^{2+}, Mg^{2+}, Fe^{2+}, Na^+, K^+, Cl^-. The value of h_{eff}/h_0 for these BECs would be rather large being in the range $10^{12} - 10^{15}$. In this case h_{eff} can be identified as gravitational Planck constant h_{gr} assignable to the magnetic flux tubes mediating gravitational interaction [8, 5, 15, 14] [22]. This would guarantee that cyclotron energies proportional to h_{eff} in endogenous magnetic field $B_{end} = 2/5B_E$, where $B_E = .5$ Gauss is the magnetic field of Earth, are above thermal energy at physiological temperature so that dark cyclotron photons can have biological effects.

2. Hydrogen bonds are central for the chemistry of water and living matter. The atoms able to form hydrogen bonds (O,N,...) possess so called lonely electron pair meaning that neither electron belongs to a valence bond.

   A possible TGD picture would be following. Hydrogen bond can be assigned with magnetic flux tube at which there is a delocalized proton, which can be also dark ($h_{eff} = n \times h_0 > h$). The lonely electron pair forms a Cooper pair. The electrons of the Cooper pair are at the members of a flux tube pair. Flux tubes are parallel but magnetic fluxes are in opposite directions if Cooper pair has spin 0. Spin 1 would correspond to fluxes in the same direction. Hydrogen bonds and their scaled up (by $h_{eff}/h_0 = n$) dark versions would correspond to flux tube pairs.

   The physics of water is plagued by anomalies. It has become recently clear that water must involve two phases. In TGD framework [25] water would have dark fraction involving dark flux tubes carrying dark protons and electrons and this would allow to understand the anomalies. Intriguingly, the anomalies are strongest at physiological temperature.

3. The basic mechanism behind cancer could be following. Free radicals steal electrons and this leads to the destruction of quantum coherence as electronic Cooper pairs are destroyed and superconductivity is lost. $h_{eff}/h_0 = n$ is reduced. This number can be regarded as a kind of IQ assignable to flux tube and one could speak about intelligence characterizing flux tube network. More precise interpretation is that the higher the value of h_{eff}/h_0 is, the higher the ability to generate conscious information is. System can also destroy information: in quantum ethics this means doing something evil!

   Remark: A little additional comment, which might irritate physicalist. TGD inspired theory of consciousness [20] suggests strongly the emergence of ethics at fundamental quantum level. Quantum ethics is simple and universal: doing good is to increase the conscious information of the Universe about itself. This conforms with the fact that doing evil forces secrecy and the Universe loses conscious information.

   The networks formed by molecules connected by flux tubes serving as correlates for quantum entanglement decay as the Planck constant at flux tubes becomes normal and they reconnect to form short loops. The community of molecules/cells decomposes into individuals, whose basic purpose degenerates to replication. Cancer is the outcome.

4. The general picture could be that the value of h_{eff}:n is reduced due to the transitions $h_{eff,i} \rightarrow h_{eff,f} < h_{eff,i}$ induced by the free radical stealing electrons. It is quite possible that the valence electron of free radical is dark.

5. What could happen in the stealing of electron? The valence electron of carcinogen (say free radical) must be dark in order that it gets on the flux tube at which Cooper pair is. Electron could be kind of Troyan horse getting to the flux tube associated with the hydrogen bond and then would react with Cooper pair splitting it and the resulting pair of electrons would consist of ordinary ordinary electrons.

2.2 Basic observation

The starting point is a reaction, in which the irradiation of carcinogen produces radiation with higher photon energy. In the example consider the incoming photon has wavelength $\lambda_i = 380$ nm and energy a $E_i = 3.27$ eV, which is just at the border 3.26 eV of violet and ultraviolet. The outgoing wavelength is $\lambda_f = 218$ n, and the corresponding energy is $E_f = 5.69$ eV and therefore in UV. As such this photon does not cause harm to say DNA.

I understand this kind of reaction is rather generally occurring for carcinogens and toxins. This suggests that the action of toxins and carcinogens is universal and relies on mechanism not depending strong the molecule considered. Understanding this on basis of standard chemistry is challenging.

I also understand that the energy 3.27 eV is special in biology and might relate to the communication between cells and that carcinogenic action somehow spoils this communications. It is also known that the emission of bio-photons in presence of carcinogen increases. If these photons are actually dark photons then dark photon BE condensate could be lost in the process and lead to a reduction of quantum coherence.

2.3 Possible detailed models for the observations of Popp

TGD based model for bio-catalysis assumes that catalyst and substrate are connected by flux tube or flux tube pair and that one can associate to this object a resonance frequency. One can ask whether carcinogen could act like catalyst.

2.3.1 Dark valence electrons behave like electrons of dark hydrogen atom

What could happen in the above process?

1. What looks strange is that the energy of final state photon is higher than initial state photon. Naively one would expect just the opposite.

   Could it be that the atom in the initial state is - in some sense not necessarily possible in standard atomic physics - in an excited state and the absorption of incoming photons makes it even more excited state. In the final state atoms returns to ground state in some sense - not necessary that of standard atomic physics. This is like jumping upwards from balcony and dropping down.

2. Electronic excitation energies for atoms must be in question. The energy scale is however too small for the transitions of hydrogen atom and even more so for those of heavier atoms. The ground state binding energy of hydrogen atom is 13.6 eV. For other atoms the energies of inner electrons are proportional Z_{eff}^2, where Z_{eff} is the effective charge of nucleus, which is screened by electrons in full shells so that Z_{eff} is considerably reduced for valence electrons.

3. How could one understand the universality? Suppose that an unpaired valence electron is in question and that it is dark. For any atom dark valence electron has orbital radius scaled up by factor $(h_{eff}/h)^2 = (n/6)^2$ so that dark valence electron sees effective nuclear charge Z_{eff}=1 and behaves like an electron of hydrogen atoms apart from small corrections coming from the mass of the nucleus! In the sequel I will call any atom with one dark valence electron (or possibly even several of them) dark hydrogen atom.

4. One can therefore assume that one has effectively transitions of dark electron of hydrogen with $h_{eff}/h = n/6 > 1$. The binding energy scale would be reduced by a factor $(h/h_{eff})^2 = (n_0/n)^2 = (6/n)^2$.

   Remark: The assumption $h = n_0 \times h_0$ raises of course bewilderment. It is however quite possible that h is not the minimal value of h_{eff}. In fact, the experiments of Randel Mills suggest $h = 6h_0$ [18]. Mills observed that hydrogen can have states for which binding energy scale is larger than normally: the would correspond to $h_{eff} = nh_0$, $n < 6$.

Remark: Recall that carcinogens are free radicals with un-paired valence electrons. These valence electrons would be dark.

2.3.2 Model I

What could be the simplest model for the reaction considered? The valence electron of dark hydrogen have spin and in ground state it could be in $n_P = 1$ tilassa (n_P is principal quantum number usually denoted by n). As it absorbs photon it can go to $n_P = 1$ state with larger value of n. One could imagine a two step process

$$(n_1, n_P) = (7,1) \rightarrow (n_2 = 8,1) \rightarrow (n_3 = 6,1) \ .$$

Could the incoming and outgoing energies be identified as energies for the transitions involved. The energies are 2.43 eV ja 5.95 eV. The actual values are 3.27 eV ja 5.69 eV. I have not found better fit so that Model I fails.

2.3.3 Model II

Let us assume a lonely dark valence electron seeing the atom effectively as hydrogen. The key observation is that $h_{eff}/h = 2$ corresponds to $n = 12 = 2 \times n_0 = 12$ with ionization energy $E_I(n = 12) = 3.4$ eV. This is not far from $E_{12} = 3.27$ eV.

Could an almost ionization from the $n = 12$ ground state with $n_P = 1$ to state $n_P = m$ occur and be followed to a state with $n < 12$ state, possibly ground state with $n_P = 1$ with an emission of photon with energy $E_{23} = 5.69$ eV $> E_{12} = 3.27$ eV? One would have $(n_i = 12, 1) \rightarrow (n_i = 12, m) \rightarrow (n_f, 1)$.

1. It is easy to see that one can have only $n_f = 9$ giving $h_{eff}/h = n_f/n_0 = n_f/6 = 3/2$. This would give ionization energy $E_I(i, n = 9) = 6.0$ eV.

2. One should have $E_{23}/E_I(n = 9) = E_{12}/E_I(n = 12) = 3.27/3.4 = .96$. The ratio of excitation energy and ground state energy would be same for initial and final state. The transition to the state $n_P = 5$ predicts $r = E_{12}/E_I(n = 12)1 - 1/25 == .96$. The prediction is correct.

   For the final state photon the prediction would be $E_{23} = (3.27/3.4) \times 6 \text{ eV} = 5.77 \text{ eV}$. The actual value is 5.69 eV. The error of the prediction is about 2 per cent.

Notice that the dark hydrogen model is extremely general and explains why so many carcinogens have this same signature. One must however notice that the orbital radius of the dark electron must be larger than that of other electrons for a screening to unit charge to take place. In the earlier applications I have assumed that the principal quantum number n_P dark valence electron is not smaller than that for the valence electrons of the ordinary atom. Also the weaker condition $n_P n > k$, where k is the principal quantum number for valence electrons, guarantees this. For hydrogen atom the condition gives no constraints but for k:th row of the periodic table one must have $n_P n > k$ (or even $n_P \geq k$). In the above model $n_P = 1$ allows only hydrogen atom. Only H and C atoms are present in carbohydrates and C has no lonely valence electrons to that the condition is automatically satisfied for them.

2.4 A model for the observations of Veljkovic

There is also an article of Veljkovic about carcinogens [3] . The article tells that the wavelength range is 206-248 nm: this would correspond to the energy range 6.1-5.0 eV in UV. On the other hand, it is noticed that the most carcinogenic wavelength range is 232-278 nm, which would correspond to the energy range 5.3-4.5 eV in UV. It would seem to me that there is a mistake in the article of Veljkovic: the upper end for wavelength range should be either 248 nm or 278 nm for both ranges. Could it be that the maximal wavelength range is 206-278 nm? TGD based model supports this interpretation as will be found.

In the first table of the article 4 absorption wavelengths have been listed for the molecules appearing in it and on basis of the summary only the lowest wavelengths can be carcinogenic.

Veljkovic does not mention the wavelength 380 nm. This suggests that this wavelength is not carcinogenic as such. On basis of what has been said the transition

$$(n = 12, 380\ nm) + X \rightarrow (n = 9, 218\ nm) + X$$

would take place. X could be a atom in bio-molecule or in carcinogen as was assumed. It is enough that dark valence electron is in question. This process would transform dark $n = 12$ photon to dark $n = 9$ photon. $(n = 12, n_P = 1)$ dark electron would go to the intermediate state $(n = 12, n_P = 5)$ and from it to $(n = 9, n_P = 1)$ dark valence electron. The reduction of h_{eff}/h_0 would mean a reduction of "biological IQ" for both dark photons and dark electrons. Could this be enough for carcinogenic effect?

One could argue, that the any molecule containing an atom with dark $n = 12$ valence electron makes it carcinogenic. This cannot be true. Carcinogen must have some additional property. Could it be that $(n = 9, 218\ nm)$ dark photons transforms to bio-photon, which is absorbed by the an ordinary electron of carcinogen, so that biological IQ is reduced further. The transformation to ordinary photon followed by absorption would be single quantum process. Note that the absorber could be also second carcinogen atom for which the absorbing valence electron is ordinary. Carcinogenicity would follow from the existence of an ordinary electronic state, which can be excited by the photon produced $(n = 12, 380\ nm) + X \rightarrow (n = 9, 218\ nm) + X$.

Assume that the transitions are those of ordinary electrons of the carcinogen. For some value of $h_{eff}/h_0 = n$ the energy range 5.0-6.1 eV could correspond to spectral lines for dark transitions of some kind creating the absorbed photons transforming to bio-photons in absorption. One can imagine two options.

Option I: The spectral lines could coincide for those for the transitions of dark hydrogen from excited state to ground state or excited state. This model turns out to be too simple to explain the observations of Veljkovic.

Option II: The spectral lines could co-incide for to those for Popp's transitions $(n = 12, n_P = 1) \rightarrow (n = 12, n_{P,i} > 1) \rightarrow n(n = 9, n_{P,f} \geq 1)$. This model can explain the observations of Veljkovic satisfactorily and suggests also a possible interpretation for the metabolic energy quantum and Coulomb energy assignable to the membrane potential. Note that only dark valence electron of hydrogen atom can be considered and this is the only possibility for hydrocarbons.

2.4.1 Option I

The transition energies of dark hydrogen characterized by n are given by

$$\frac{\Delta E}{E_H} = (\frac{6}{n})^2[\frac{1}{n_{P,f}^2} - \frac{1}{n_{P,i}^2}] \ . \tag{2.1}$$

The simplest option is that the transition takes place to the ground state with $n_{P,f} = 1$.

For what value of $h_{eff}/h_0 = n$ energy range 5.0-6.1 eV could correspond to the spectral lines of dark hydrogen? Ionization energy for dark hydrogen gives the largest energy and it should be around $E_{max} = 6.1$ eV. If ionization does not take place, photon energy is lower and could correspond to energies in the range 5.0-6.1 eV. With these assumptions one obtains

$$E_H(n) = E_H \times (\frac{h}{h_{eff}})^2 = E_H \times (\frac{6}{n})^2 = E_{max} \ , \quad E_{max} = 6.1 \text{ eV} \ , \quad E_H = 13.6 \text{ eV} \ . \tag{2.2}$$

This gives $n^2 = 80.26$ so that n is very near to $n = 9$ and $h_{eff} = 3h/2$. $n = 9$ gives upper bound $E_{max} = 6.04$ eV. Also other energies could correspond to the transitions of dark hydrogen. The transition would be of for $(n_P \rightarrow n_P = 1)$ and the energy of the emitted photon would satisfy the condition

$$\Delta E = E_{max}(1 - n_P^{-2}) = E_{min} = 5.0 \text{ eV} . \quad (2.3)$$

This would give $1/n_P^2 = 1/6$ in a reasonable approximation. This cannot be true. What if one uses as the lower bound the energy $E_{min} = 4.5$ eV, which corresponds to 278 nm. This would give $m = 2$ for $E_{max} = 6$ eV! The maximal range 206-278 nm would correspond to the emission spectrum for the transition to the ground state. Getting dark counterparts for the absorbtion energies listed by Veljkovic does not however seem probable since there is only single integer valued parameter available.

2.4.2 Option II

The energy of the photon is difference of $n_f = 9$ and $n_i = 12$ excitation energies characterized by $n_{P,f}$ ja $n_{P,i}$. The general formula for the transition energy ΔE allowing n_f and n_i to be arbitrary reads as

$$\Delta E = [\frac{1}{n_{P,f}^2}(\frac{6}{n_f})^2 - \frac{1}{n_{P,i}^2}(\frac{6}{n_i})^2]E_H . \quad (2.4)$$

For $n_i = 12$ with $h_{eff} = 2h$ and $n_f = 9$ with $h_{eff} = 3h/2$ one obtains the formula

$$\Delta E = [\frac{4}{9n_{P,f}^2} - \frac{1}{4n_{P,i}^2}]E_H , \quad E_H = 13.6 \text{ eV} . \quad (2.5)$$

Consider first the dependence of ΔE on n_i for given n_i.

1. Consider first the situation for $n_{P,f} = 1$.
   (a) ΔE is largest at the limit $n_{P,i} \to \infty$: this gives $\Delta E = (4/9)E_H = 6.04$ eV ($\lambda = 205$ nm), which corresponds to the upper bound for energies deducible from the results of Veljkovic. This energy is also largest possible since the scale of ΔE is proportional to $1/n_{P,f}^2$.
   (b) One obtains minimum of ΔE for for $n_{P,i} = 1$ as $\Delta E = 2.64$ eV ($\lambda = 469$ nm, blue). One therefore obtains has a band [2.64 ,6.04] eV of lines become dense at its UV end.
   (c) For $n_{P,i} = 2$ gives ($\Delta E = 5.19$ eV, $\lambda = 239$ nm). The wavelength is near to the lower bound of the wavelength range 232-278 nm mentioned by Veljkovic. For $n_{P,i} = 3$ one obtains ($\Delta E = 5.67$ eV, $\lambda = 219$ nm). The wavelength approaches to the limit 205 nm at the limit $n_{P,i} \to \infty$. The wave lengths are very densely spaced for large values of $n_{P,i}$ could well correspond in good enough approximation to the wavelengths near the lower boundary of the wavelength range given by Veljkovic.

2. $n_{P,f} = 2$ gives the upper bound $\Delta E = 1.5$ eV for ($n_{P,i} \to \infty$ in near infrared ($\lambda = 821$ nm). Lower bound $\Delta E = .67$ eV is obtained for $n_{P,i} = 2$. One has therefore band [.67,1.5] eV with density of lines getting dense in near infrared. Quite generally, for $n_{P,f} \geq 2$ ΔE is below UV range and arbitrary small values of ΔE are possible for large enough values of $n_{P,f}$.

3. $n_{P,f} = 3$ gives the upper bound $\Delta E \leq .67$ eV for ($n_{P,i} \to \infty$ and lower bound $\Delta E = .294$ eV for $n_{P,i} = 3$. Metabolic energy quantum with value of .5 eV is included in this range of energies and $n_{P,f} = 9$ gives $\Delta E = 4.45$ eV.

Consider next the minimal values of the energy for given $n_{P,f}$.

1. The condition $\Delta E \geq 0$ gives $n_{P,f} \leq 4n_i/3$. For $(n_{P,f}, n_{P,i}) = k(4,3)$ one has $\Delta E = 0$. For $n_{P,f}$ integer nearest to but smaller than $n_{P,f} = 4n_{P,i}/3 - 1$ one has smallest value of ΔE for given $n_{P,i}$. The following formula for ΔE for $n_{P,f} = 4n_{P,i}/3 - 1$ is true for $n_{P,i} = 3k$:

$$\Delta E_{in}(n_{P,i}) = [\frac{4}{9}\frac{1}{(4k-1)^2} - \frac{1}{36k^2}]E_H \simeq \frac{E_H}{72k^3} \simeq \frac{.19 \text{ eV}}{k^3} \text{ for } k \rightarrow \infty \ . \tag{2.6}$$

2. For $k = 1$ ($n_{P,i} = 3$) one obtains ($\Delta E = .66$ eV,$\lambda = 1879$ nm). ΔE is slightly higher than the nominal value .5 eV of the metabolic energy quantum.

3. For $k = 2$ ($n_{P,i} = 6$) one obtains $\Delta E = .065$ eV, which corresponds to a typical membrane potential.

To summarize, Popp transition energies of dark valence electrons of dark hydrogen atom might explain not only the energies listed by Veljkovic but also metabolic energy quantum and Josephson energy assignable to cell membrane in TGD based model of cell membrane as generalized Josephson junction.

2.5 Could the dark photons from Popp transitions transform to bio-photons?

Bio-photons do not seem to be produced by molecular transitions although they can induce molecular transitions about which the transitions of carcinogens would be and example. I have proposed earlier that bio-photons include dark cyclotron photons with harmonic oscillator spectrum. Spectra for several strengths of magnetic field are required to get a quasi-continuum believed to characterize bio-photons. For dark cyclotron photons also the value of $h_{eff} = h_{gr}$ would be very large [15] [22]. The photons emitted in the transitions of dark valence electrons with relatively small value of h_{eff} serve also as a candidate for dark photons transforming to bio-photons. They could be assigned to the parts of the magnetic body with relatively small size scale (say flux tubes connecting cells) unlike those with large value of h_{eff} and wavelengths even of order those of EEG photons.

Bio-photons include also visible wave length range. Do the transitions of dark hydrogen allow to cover this range? Besides the above kind of transitions reducing h_{eff}/h, one can also consider the transitions increasing it. One might argue that the transitions responsible for color vision are of latter type since negentropy increase is involved.

The following **Tables 1** and **2** describe the energies of emitted photons in processes ($n_i \rightarrow n_f$) with $n_{P,i} = 1$ in the case that they are kinematically possible. n_i and n_f are allowed to vary in the range $(9, ..., 17)$ so that transitions which either increase or reduce h_{eff}/h, or leave it unaffected, are allowed.

Remark: The condition $n_{P,i}n_i > k$, where k is principle quantum number for the valence electrons of the ordinary atom guarantees the screening to unit charge. $n_{P,i} \geq k$ assumed in earlier models would be stronger condition. Similar condition must be satisfied by $n_{P,f}$: the transitions with $n_{P,i} \leq n_{P,f}$ are always possible.

1. The rows of the tables with fixed n_i give the minimum value $n_{P,f,min}$ of $n_{P,f}$ determined by the condition that the photon energy ΔE is positive, the energy ΔE_{min} in this case, and the maximum ΔE_{max} for which final state electron is free ($n_{P,f} \rightarrow \infty$. The transitions for $n_{P,f} < n_{P,f,min}$ can occur in reversal time direction as absorption.

2. By changing the roles of n_i and n_f and of $n_{P,i} = 1$ and $n_{P,f}$, the same table gives some transition energies with final state electron in the ground state ($n_{P,f} = 1$). The table also gives minimal absorption energies ΔE_{min} *resp.* maximal absorption energies ΔE_{min} as function of n_i and $n_{P,i,max}$ *resp.* $n_{P,i,max}$. Note that the transitions for $n_{P,i} < n_{P,i,min}$ for which photon energy would be negative can occur in reversal time direction as emission.

From the tables one learns that the energies of photons in visible regions can be covered by the scaled variants of the spectra but the regions near the ends have a low density of lines.

1. The densities of the spectral lines increase towards the maximal energies $\Delta E_{max}/eV \in (1.69, 1.91, 2.18, 2.5, 2.90, 3.40, 4.05, 4.90, 6.04)$ associated with $17 \geq n_i \geq 9$). The upper ends of the frequency range for $n_i + 1$ are above the lower ends for n_i so that the ranges of energies overlap. The deviation from un-evenness can be testable someday as detection technologies develop.

2. As a rule, the spectra for the transitions reducing h_{eff} begin at $n_{P,f} = 2$ since the lowest state would correspond to negative energy. The transition can be however realized in opposite direction as as a transition increasing h_{eff}. I have added to ΔE_{min} column (fourth column) the energy of this transition in brackets.

   I have added to ΔE_{min} column (fourth column) 2 spectral lines in brackets to show where the visible part of the spectrum begins in these cases. The reader can compare the spectrum to the data given about the spectrum of visible light (see `http://tinyurl.com/q8yqea9`).

A couple of comments about the interpretation of the spectrum is in order.

1. The maximum energies for the bands intersecting visible range are $\Delta E \in (1.69, 1.91, 2.18, 2.5, 2.90, 3.40)$ labelled by $17 \geq\geq n_i \geq 12$). Note that upper end of violet is 3.26 eV and belongs to the band [2.55, 3.40] eV containing blue. Could these 6 bands becoming infinitely dense towards their upper ends correspond to the 6 color-complement color pairs red-green, blue yellow and white-black pair included? Could different values of n_i characterize color qualia? Could the ends of the bands be identified as "nominal" wavelengths for the basic colors? Note that I have constructed a model for color vision relying on the transitions of dark electrons in [23].

2. I have also suggested that music harmony could emerge at the level of fundamental physics [16, 26], in particular the model for dark genetic code [21] leads to 12-note scale. An interesting questions is whether the ratios for the frequencies associated with $\Delta E \in (1.69, 1.91, 2.18, 2.5, 2.90, 3.40)$ could correspond to simple music scale. The ratios of the energies to the smallest energy are given by $(1.00, 1.13, 1.29, 1.5, 1.72, 2.01)$. In even tempered scale with the notes of 12-note scale coming as $f_n/f_0 = 2^{n/12}$ one obtains for the pentatonic scale C,D,E,G,A,C appearing in Chinese music the frequencies ratios $(1.00, 1.12, 1.26, 1.50, 1.68, 2.00)$. The deviations are few per cent.

3 Possible general mechanisms for the action of carcinogen

In the following some general guesses for the effect of carcinogens are discussed and after that a model based on the findings of Popp and Veljkovic is discussed.

3.1 Some general ideas

Consider first some guesses.

1. The dark photons of BEC can be absorbed and reduce also reduce the value of n for dark electrons: for instance, in the above example one has $n_i = 12 \rightarrow n_f = 9$.

2. This reduction of n for catalyst and return to its original value possibly requiring metabolic energy would be the basic mechanism of bio-catalysis. It would liberate temporarily metabolic energy allowing to overcome the potential wall slowing down the reaction considered.

   Carcinogens would imitate other biomolecules in that they would have dark electrons. This might help to get into bio-molecules in this manner (consider bentzene as example). Dark lonely unpaired valence electrons would be in fundamental role. Their transitions would produce a universal spectrum playing a key role in the bio-control.

Table 1: Table represents minimal and maximal dark photon energies $\Delta E_{min}/eV$ and $\Delta E_{max}/eV$ for transitions $(n_i, n_{P,i}) \rightarrow (n_f, n_{P,f})$ in the range $n_i \in [9, 14]$. In the column for $\Delta E_{min}/eV$ numbers in brackets give for $n_f = 1$ rows the $n_{P,i} = 2, 3$ transition energies and for $n_{P,f} = 2$ rows transition energy for the reverse transition $(1, 1) \rightarrow (1, 1)$.

n_i	n_f	$n_{P,f,min}$	$\Delta E_{min}/eV$	$\Delta E_{max}/eV$
9	9	2	4.53	6.04
9	10	1	1.15 (4.82,5.50)	6.04
10	9	2	(1.15) 3.38	4.90
10	10	2	3.67	4.90
10	11	1	0.85 (3.88,4.45)	4.90
11	9	2	(2.00) 2.54	4.05
11	10	2	(0.85) 2.82	4.05
11	11	2	3.03	4.05
11	12	1	0.65 (3.20,3.67)	4.05
12	9	2	(2.64) 1.89	3.40
12	10	2	(1.50) 2.18	3.40
12	11	2	(0.65) 2.39	3.40
12	12	2	2.55	3.40
12	13	1	0.50 (2.68,3.08)	3.40
13	9	2	(3.15) 1.39	2.90
13	10	2	(2.00) 1.67	2.90
13	11	2	(1.15) 1.89	2.90
13	12	2	(0.50) 2.05	2.90
13	13	2	2.17	2.90
13	14	1	0.40 (2.27,2.62)	2.90
14	9	2	(3.55) 0.99	2.50
14	10	2	(2.40) 1.27	2.50
14	11	2	(1.55) 1.49	2.50
14	12	2	(0.90)1.65	2.50
14	13	2	(0.40) 1.77	2.50
14	14	2	1.87	2.50
14	15	1	0.32 (1.95,2.26)	2.50

Table 2: Table represents minimal and maximal dark photon energies $\Delta E_{min}/eV$ and $\Delta E_{max}/eV$ for transitions $(n_i, n_{P,i}) \rightarrow (n_f, n_{P,f})$ in the range $n_i \in [15, 17]$. In the column for $\Delta E_{min}/eV$ numbers in brackets give for $n_f = 1$ rows the $n_{P,i} = 2, 3$ transition energies and for $n_{P,f} = 2$ rows transition energy for the reverse transition $(1, 1) \rightarrow (1, 1)$.

n_i	n_f	$n_{P,f,min}$	$\Delta E_{min}/eV$	$\Delta E_{max}/eV$
15	9	2	(?) 0.66	2.18
15	10	2	((2.72) 0.95	2.18
15	11	2	(1.87) 1.16	2.18
15	12	2	(1.22) 1.33	2.18
15	13	2	(0.72) 1.45	2.18
15	14	2	(0.32) 1.55	2.18
15	15	2	1.63	2.18
15	16	1	0.26 (1.70,1.96)	2.18
16	9	2	(4.13) 0.40	1.91
16	10	2	(2.98) 0.69	1.91
16	11	2	(2.13) 0.90	1.91
16	12	2	(1.49) 16	1.91
16	13	2	(0.98) 1.19	1.91
16	14	2	(0.59) 1.29	1.91
16	15	2	(0.26) 1.37	1.91
16	16	2	1.43	1.91
16	17	1	0.22 (1.40,1.72)	1.91
17	9	2	(4.35) 0.18	1.69
17	10	2	(3.20) 0.47	1.69
17	11	2	(2.35) 0.68	1.69
17	12	2	(1.71) 0.84	1.69
17	13	2	1.20) 0.97	1.69
17	14	2	(0.80) 17	1.69
17	15	2	(0.48) 1.15	1.69
17	16	2	(0.22) 1.22	1.69
17	17	2	1.27	1.69
17	18	1	0.18 (1.32,1.53)	1.69

3. If 3.27 eV:n photons emerge t $n = 12$ BEC assignable to organism, the presence of carcinogen would lead to a loss of the BEC and production of bio-photons.

If this is the case, the spectra for the transitions $(n_i, n_{P,i} \rightarrow (n_f, n_{P,f})$ of dark hydrogen atom would define the central frequencies and key energies of bio-control. There would be infinite number of these corresponding to all transitions $(n_1, n_{P,1}) \rightarrow (n_2, n_{P_2})$. Energy difference and at the same time the spectrum of biologically important photons would contain the transition energies of dark hydrogen atom:

$$E((n_i, n_{P,i} \rightarrow (n_f, n_{P,f}) = \frac{1}{n^2}[\frac{1}{n_f^2}\frac{1}{n_{P,f}^2} - \frac{1}{n_i^2}\frac{1}{n_{P,i}^2}] \times E_I(H) \ , \quad E_H = 13.6 \ eV \ . \tag{3.1}$$

One can say, that these spectra produce a fractal, since they are obtained from each other by scaling using rational number. Here the value of n can be such that the energies are in visible and UV range corresponding to the energy spectrum of bio-photons. The dynamics of living matter would be universal, which conforms with quantum criticality.

One could think that if molecule has in its ordinary spectrum a line coinciding with some energy in above spectrum, the molecule defines a potential carcinogen. All atoms with un-paired valence electron, which can be dark would be potential parts of carcinogen. Some additional condition must be satisfied for a molecule to be a carcinogen: the existence of ordinary transition with energy in the dark photon spectrum could be this condition. There are also other frequency spectra such as cyclotron transitions and also these could couple to carcinogens.

3.2 A proposal for the carcinogenic mechanism inspired by the observations of Popp and Veljkovic

This picture encourages to consider a rather simple mechanism for cancer as a loss of quantum coherence due to the decay of Bose-Einstein condensate of dark photons caused by the presence of carcinogen molecules. Also super conductivity possibly associated with dark valence electrons might be lost. Carcinogen would absorb the $n = 9$ dark photons ($\lambda = 218$ nm) generated from $n = 12$ dark photons (for instance for $\lambda = 380$ nm) by Popp mechanism.

Dark photon, call it A, would transform with certain rate $k_{A \rightarrow B}$ to ordinary photon (bio-photon). Bio-photon would transform with rate $k_{B \rightarrow A}$ to dark photon. Carcinogen molecule would absorb bio-photons B with rate k_C. The situation is analogous to a chemical reaction in which second components leaks out from the system by reacting with a third component, whose concentration is assumed to be large. The outcome is that both A and B approach to zero and BEC is lost.

For the densities of photons obtains the equations

$$\begin{aligned} \frac{dA}{dt} &= k_{B \rightarrow A}B - k_{A \rightarrow B}A \ , \\ \frac{dB}{dt} &= -k_{B \rightarrow A}B + k_{A \rightarrow B}A - k_C B \ . \end{aligned} \tag{3.2}$$

The equations are linear and the solution is sum of two exponent terms with rather free coefficients (A and B must be positive).

The general form for the equations is

$$\begin{aligned} \frac{dA}{dt} &= k_1 B - k_2 A \ , \\ \frac{dB}{dt} &= -k_3 B + k_2 A \ . \end{aligned} \tag{3.3}$$

One has

$$k_1 = k_{B\rightarrow A} \ , \quad k_2 = k_{A\rightarrow B} \quad k_3 = k_{B\rightarrow A} + k_C \ . \tag{3.4}$$

One has $k_3 > k_1$ since B is absorbed by carcinogen.

By using the ansatz

$$A = A_0 exp(-kt) \ , \quad B = B_0 exp(-kt) \ . \tag{3.5}$$

one obtains a homogenous linear group of two equations and the solutions for k are determined by the vanishing of the determinant of the matrix defining the group

$$k_\pm = \frac{k_1+k_3}{2} \pm \frac{1}{2}\sqrt{(k_3+k_1)^2 - 4(k_3+k_2)k_1} \ . \tag{3.6}$$

The general solution is of the form

$$\begin{pmatrix} A \\ B \end{pmatrix} = \sum_\pm a_\pm exp(-k_\pm t) \begin{pmatrix} \frac{k_2}{k_\pm + k_1} \\ 1 \end{pmatrix} \ . \tag{3.7}$$

Both A and B approach zero with an exponential rate.

4 Appendix: Number theoretical characterization of the photon spectrum from dark valence electron transitions

The spectrum for the lines of dark photons from the hydrogen-like transitions of dark valence electron can be characterized number theoretically. The reason is that given transition energy is characterized by a pair (k_i, k_f) of products integers $k_i = n_i n_{P,i}$ and $k_f = n_f n_{P,f}$ as

$$\frac{\Delta E}{E_H} = \frac{1}{k_i^2} - \frac{1}{k_f^2} \ . \tag{4.1}$$

For given k_i *resp.* k_f all its decompositions to a product of integers define one possible initial *resp.* final state. The spectral density is sum of energy conserving delta functions each multiplied by the number of transitions with the energy consider. This number is proportional to the product $N(k_i)N(k_f)$ for the numbers of these decompositions for k_i and k_f. The spectral density function has therefore a large value when both k_i and k_f have large number of factors.

Could the photons produced in this kind of transitions could be of special physical and biological significance? This could be the case if the number of allowed pairs (n_i, n_f) and $(n_{P,i}, n_{P,f})$ is large enough. Whether this could be the case is an open question. In any case it is interesting to look what this would imply.

One has always the decompositions $(n = 1, n_P = k)$ and $(n = k, n_P = 1)$ and for prime values of k only these decompositions exist. For non-prime values of k there are also decompositions to a product of integers different from k and 1. The number $N(k)$ of factorizations of k into a product of two integers is given by the number of different factors of k. Elementary argument showing that the number of decompositions of p^r equals to $N(p^r) = r + 1$ shows that $N(k)$ is obtained from the prime decomposition $k = \prod p_i^{r_i}$ of k as

$$N(k) = \prod_i (r_i + 1) \ , \quad k = \prod_i p_i^{r_i} \ . \tag{4.2}$$

For numbers k_i having large number of different factors the number of product decompositions is large. For prime values of k_i there are only two compositions. For instance, factorial $k = r! = 1 \times 2... \times r$ the number of decompositions is large. Powers $k = p^r$ have $N(k) = r + 1$ decompositions. Perfect numbers $P = M_p 2^{p-1}$ ($M_p = 2^p - 1$) have large number of composition due to the large power of 2 involved.

An interesting question is, for which kind of integers the number of factors divided by integer is maximal.It is known that $N(n)$ satisfied the inequality $N(n) \leq 2^{1.5379 log(n)/log(log(n))}$ and that equation holds true for $N = 6,983,776,800$ (see `http://tinyurl.com/yar9kdfd` and `http://tinyurl.com/y7nvfce5`). I do not know whether the equation is true for some other integers. Just for fun one can look the frequency and period associated with the ground state energy of hydrogen atom with $h_{eff} = Nh_0$ assuming $h = 6h_0$. The frequency is $f = (6/N)^2(E_H/eV) \times (3/1.24)10^{14}$ Hz giving period $T = 1/f = 187.4$ h or 7.8 days, with day=24 h.

Assuming that all transitions have the same probability to appear (an assumption very probably non-realistic), one can write the spectral density function as the density of states per energy as a sum of energy conserving delta functions multiplied by the number $N(k_i)N(k_f)$ of transition with this energy

$$\frac{dN}{dE} = \sum_{k_i,k_f} N(k_i)N(k_f)\delta(E - E_{k_i \to k_f}) \ . \tag{4.3}$$

Therefore the pairs (k_i, k_f) with both integers having large number of factors could be of special interest. In a more realistic treatment each delta function contains an additional weight factor telling the probability for the particular transition to occur.

Acknowledgements: I am grateful for Tommi Ullgren for informing me about the work of Popp and Veljkovic related to the interaction of carcinogens with UV light.

References

[1] Preskill J et al. Holographic quantum error-correcting codes: Toy models for the bulk/boundary correspondence. Available at: `http://arxiv.org/pdf/1503.06237.pdf`, 2015.

[2] Popp FA. Electronic structure and carcinogenic activity of 3,4-benzopyrene and 1,2-benzopyrene. *Z Naturforsch B.*, 27(7):850–863, 1972.

[3] Veljkovic V and Lalovic DI. Correlation between the carcinogenicity of organic substances and their spectral characteristics. *Experientia*, 34(10):1342–1343, 1978.

[4] Pitkänen M. Magnetic Sensory Canvas Hypothesis. In *TGD and EEG*. Online book. Available at: `http://www.tgdtheory.fi/public_html/tgdeeg/tgdeeg.html#mec`, 2006.

[5] Pitkänen M. Quantum Astrophysics. In *Physics in Many-Sheeted Space-Time*. Online book. Available at: `http://www.tgdtheory.fi/public_html/tgdclass/tgdclass.html#qastro`, 2006.

[6] Pitkänen M. Quantum Model for Bio-Superconductivity: I. In *TGD and EEG*. Online book. Available at: `http://www.tgdtheory.fi/public_html/tgdeeg/tgdeeg.html#biosupercondI`, 2006.

[7] Pitkänen M. Quantum Model for Bio-Superconductivity: II. In *TGD and EEG*. Online book. Available at: `http://www.tgdtheory.fi/public_html/tgdeeg/tgdeeg.html#biosupercondII`, 2006.

[8] Pitkänen M. TGD and Astrophysics. In *Physics in Many-Sheeted Space-Time*. Online book. Available at: `http://www.tgdtheory.fi/public_html/tgdclass/tgdclass.html#astro`, 2006.

[9] Pitkänen M. *TGD and EEG*. Online book. Available at: `http://www.tgdtheory.fi/public_html/tgdeeg/tgdeeg.html`, 2006.

[10] Pitkänen M. Quantum Mind, Magnetic Body, and Biological Body. In *TGD based view about living matter and remote mental interactions.* Online book. Available at: http://www.tgdtheory.fi/public_html/tgdlian/tgdlian.html#lianPB, 2012.

[11] Pitkänen M. *TGD Based View About Living Matter and Remote Mental Interactions.* Online book. Available at: http://www.tgdtheory.fi/public_html/tgdlian/tgdlian.html, 2012.

[12] Pitkänen M. Are dark photons behind biophotons. In *TGD based view about living matter and remote mental interactions.* Online book. Available at: http://www.tgdtheory.fi/public_html/tgdlian/tgdlian.html#biophotonslian, 2013.

[13] Pitkänen M. Comments on the recent experiments by the group of Michael Persinger. In *TGD based view about living matter and remote mental interactions.* Online book. Available at: http://www.tgdtheory.fi/public_html/tgdlian/tgdlian.html#persconsc, 2013.

[14] Pitkänen M. Criticality and dark matter. In *Hyper-finite Factors and Dark Matter Hierarchy.* Online book. Available at: http://www.tgdtheory.fi/public_html/neuplanck/neuplanck.html#qcritdark, 2014.

[15] Pitkänen M. Quantum gravity, dark matter, and prebiotic evolution. In *Genes and Memes.* Online book. Available at: http://www.tgdtheory.fi/public_html/genememe/genememe.html#hgrprebio, 2014.

[16] Pitkänen M. Geometric theory of harmony. Available at: http://tgdtheory.fi/public_html/articles/harmonytheory.pdf, 2014.

[17] Pitkänen M. Holography and Quantum Error Correcting Codes: TGD View. Available at: http://tgdtheory.fi/public_html/articles/tensornet.pdf, 2016.

[18] Pitkänen M. Hydrinos again. Available at: http://tgdtheory.fi/public_html/articles/Millsagain.pdf, 2016.

[19] Pitkänen M. Artificial Intelligence, Natural Intelligence, and TGD. Available at: http://tgdtheory.fi/public_html/articles/AITGD.pdf, 2017.

[20] Pitkänen M. Re-examination of the basic notions of TGD inspired theory of consciousness. Available at: http://tgdtheory.fi/public_html/articles/conscrit.pdf, 2017.

[21] Pitkänen M. About dark variants of DNA, RNA, and amino-acids. Available at: http://tgdtheory.fi/public_html/articles/darkvariants.pdf, 2018.

[22] Pitkänen M. About the physical interpretation of the velocity parameter in the formula for the gravitational Planck constant . Available at: http://tgdtheory.fi/public_html/articles/vzero.pdf, 2018.

[23] Pitkänen M. Dark valence electrons and color vision. Available at: http://tgdtheory.fi/public_html/articles/colorvision.pdf, 2018.

[24] Pitkänen M. Emotions as sensory percepts about the state of magnetic body? Available at: http://tgdtheory.fi/public_html/articles/emotions.pdf, 2018.

[25] Pitkänen M. Maxwells lever rule and expansion of water in freezing: two poorly understood phenomena. Available at: http://tgdtheory.fi/public_html/articles/leverule.pdf, 2018.

[26] Pitkänen M. New results in the model of bio-harmony. Available at: http://tgdtheory.fi/public_html/articles/harmonynew.pdf, 2018.

Essay

Thoughts on Modification of Bio-harmony

Matti Pitkänen [1]

Abstract

I have constructed a model of bio-harmony as a fusion of 3 icosahedral harmonies and tetrahedral harmony. The icosahedral harmonies are defined by Hamiltonian cycles at icosahedron going through every vertex of the icosahedron and therefore assigning to each triangular face an allowed 3-chord of the harmony. The fascinating outcome is that the model can reproduces genetic code. The model for how one can understand how 12-note scale can represent 64 genetic codons has the basic property that each note belongs to 16 chords. The reason is that there are 3 disjoint sets of notes and given 3-chord is obtained by taking 1 note from each set. For bio-harmony obtained as union of 3 icosahedral harmonies and tetrahedral harmony note typically belongs to 15 chords. The representation in terms of frequencies however requires 16 chords per note. Consistency a modification of the model of icosahedral harmony. The necessity to introduce tetrahedron for one of the 3 fused harmonies is indeed an ugly looking feature of the model. The question is whether one of the harmonies could be replaced with some other harmony with 12 notes and 24 chords. If this would work one would have 64 chords equal to the number of genetic codons and 5+5+6 =16 chords per note. One can imagine toric variants of harmonies realized in terms of Hamiltonian cycles and one indeed obtains a toric harmony with 12 notes and 24 3-chords. Bio-harmony could correspond to the fusion of 2 icosahedral harmonies with 20 chords and toric harmony with 24 chords having therefore 64 chords. Whether the predictions for the numbers of codons coding for given amino-acids come out correctly for some choices of Hamiltonian cycles is still unclear.

Keywords: Bio-harmony, modification, genetic code, TGD framework.

1 Introduction

I have developed a rather detailed model of bio-harmony as a fusion of 3 icosahedral harmonies and tetrahedral harmony [3, 4](see http://tinyurl.com/yad4tqwl and http://tinyurl.com/y8njuctq). The icosahedral harmonies are defined by Hamiltonian cycles at icosahedron going through every vertex of the icosahedron and therefore assigning to each triangular face an allowed 3-chord of the harmony. The surprising outcome is that the model can reproduces genetic code.

The model for how one can understand how 12-note scale can represent 64 genetic codons has the basic property that each note belongs to 16 chords. The reason is that there are 3 disjoint sets of notes and given 3-chord is obtained by taking 1 note from each set. For bio-harmony obtained as union of 3 icosahedral harmonies and tetrahedral harmony note typically belongs to 15 chords. The representation in terms of frequencies requires 16 chords per note.

If one wants consistency one must somehow modify the model of icosahedral harmony. The necessity to introduce tetrahedron for one of the 3 fused harmonies is indeed an ugly looking feature of the model. The question is whether one of the harmonies could be replaced with some other harmony with 12 notes and 24 chords. If this would work one would have 64 chords equal to the number of genetic codons and 5+5+6 =16 chords per note. The addition of tetrahedron would not be needed.

One can imagine toric variants of harmonies realized in terms of Hamiltonian cycles and one indeed obtains a toric harmony with 12 notes and 24 3-chords. Bio-harmony could correspond to the fusion of 2 icosahedral harmonies with 20 chords and toric harmony with 24 chords having therefore 64 chords. Whether the predictions for the numbers of codons coding for given amino-acids come out correctly for

[1]Correspondence: Matti Pitkänen http://tgdtheory.fi/. Address: Rinnekatu 2-4 8A, 03620, Karkkila, Finland. Email: matpitka6@gmail.com.

some choices of Hamiltonian cycles is still unclear. This would require an explicit construction of toric Hamiltonian cycles.

Before discussing the possible role of toric harmonies some previous results will be summarized.

1.1 Icosahedral bio-harmonies

The model of bio-harmony [3] starts from a model for music harmony as a Hamiltonian cycle at icosahedron having 12 vertices identified as 12 notes and 20 triangular faces defining the allowed chords of the harmony. The identification is determined by a Hamiltonian cycle going once through each vertex of icosahedron and consisting of edges of the icosahedral tesselation of sphere (analog of lattice): each edge corresponds to quint that is scaling of the frequency of the note by factor $3/2$ (or by factor $2^{7/12}$ in well-tempered scale). This identification assigns to each triangle of the icosahedron a 3-chord. The 20 faces of icosahedron define therefore the allowed 3-chords of the harmony. There exists quite a large number of icosahedral Hamiltonian cycles and thus harmonies.

The fact that the number of chords is 20 - the number of amino-acids - leads to the question whether one might somehow understand genetic code and 64 DNA codons in this framework. By combining 3 icosahedral harmonies with different symmetry groups identified as subgroups of the icosahedral group, one obtains harmonies with 60 3-chords.

The DNA codons coding for given amino-acid are identified as triangles (3-chords) at the orbit of triangle representing the amino-acid under the symmetry group of the Hamiltonian cycle. The predictions for the numbers of DNAs coding given amino-acid are highly suggestive for the vertebrate genetic code.

By gluing to the icosahedron tetrahedron along common face one obtains 4 more codons and two slightly different codes are the outcome. Also the 2 amino-acids Pyl and Sec can be understood. One can also regard the tetrahedral 4 chord harmony as additional harmony so that one would have fusion of four harmonies. One can of course criticize the addition of tetrahedron as a dirty trick to get genetic code.

The explicit study of the chords of bio-harmony however shows that the chords do not contain the 3-chords of the standard harmonies familiar from classical music (say major and minor scale and corresponding chords). Garage band experimentation with random sequences of chords requiring conservability that two subsequent chords have at least one common note however shows that these harmonies are - at least to my opinion - aesthetically feasible although somewhat boring.

1.2 Explanation for the number 12 of notes of 12-note scale

One also ends up to an argument explaining the number 12 for the notes of the 12-note scale [3]. There is also second representation of genetic code provided by dark proton triplets. The dark proton triplets representing dark genetic codons are in one-one correspondence with ordinary DNA codons. Also amino-acids, RNA and tRNA have analogs as states of 3 dark protons. The number of tRNAs is predicted to be 40.

The dark codons represent entangled states of protons and one cannot decompose them into a product state. The only manner to assign to the 3-chord representing the triplet ordinary DNA codon such that each letter in {A,T,C,G} corresponds to a frequency is to assume that the frequency depends on the position of the letter in the codon. One has altogether $3 \times 4 = 12$ frequencies corresponding to 3 positions for given letter selected from four letters.

Without additional conditions any decomposition of 12 notes of the scale to 3 disjoint groups of 4 notes is possible and possible chords are obtained by choosing one note from each group. The most symmetric choice assigns to the 4 letters the notes $\{C, C\sharp, D, D\sharp\}$ in the first position, $\{E, F, F\sharp, G\}$ in the second position, and $\{G\sharp, A, B\flat, B\}$ in the third position. The codons of type XXX would correspond to $CEG\sharp$ or its transpose. One can transpose this proposal and there are 4 non-quivalent transposes, which could be seen as analogs of music keys.

Remark: $CEG\sharp$ between C-major and A-minor very often finishes finnish tango: something neither sad nor glad!

One can look what kind of chords one obtains.

1. Chords containing notes associated with the same position in codon are not possible.

2. Given note belongs to 6 chords. In the icosahedral harmony with 20 chords given note belongs to 5 chords (there are 5 triangles containing given vertex). Therefore the harmony in question cannot be equivalent with 20-chord icosahedral harmony. Neither can the bio-harmony with 64 chords satisfy the condition that given note is contained by 6 3-chords.

3. First and second notes of the chords are separated by at least major third as also those second and third notes. The chords satisfy however octave equivalence so that the distance between the first and third notes can be smaller - even half step - and one finds that one can get the basic chords A-minor scale: Am, Dm, E7, and also G and F. Also the basic chords of F-major scale can be represented. Also the transposes of these scales by 2 whole steps can be represented so that one obtains A_m, $C\sharp_m$, F_m and corresponding major scales. These harmonies could allow the harmonies of classical and popular music.

These observations encourage to ask whether a representation of the new harmonies as Hamiltonian cycles of some tesselation could exist. The tesselation should be such that 6 triangles meet at given vertex. Triangular tesselation of torus having interpretation in terms of a planar parallelogram (or perhaps more general planar region) with edges at the boundary suitable identified to obtain torus topology seems to be the natural option. Clearly this region would correspond to a planar lattice with periodic boundary conditions.

2 Is it possible to have toric harmonies?

The basic question is whether one can have a representation of the new candidate for harmonies in terms of a tesselation of torus having $V = 12$ vertices and $F = 20$ triangular faces. The reading of the article "*Equivelar maps on the torus*" [1] (see `http://tinyurl.com/ya6g9kwe`) discussing toric tesselations makes clear that this is impossible. One however have $(V, F) = (12, 24)$ (see `http://tinyurl.com/y7xfromc`). A rather promising realization of the genetic code in terms of bio-harmony would be as a fusion of two icosahedral harmonies and toric harmony with $(V, F) = (12, 24)$. This in principle allows also to have 24 3-chords which can realize classical harmony (major/minor scale).

1. The local properties of the tesselations for any topology are characterized by a pair (m, n) of positive integers. m is the number of edges meeting in given vertex (valence) and n is the number of edges and vertices for the face. Now one has $(m, n) = (6, 3)$. The dual of this tesselation is hexagonal tesselation $(m, n) = (3, 6)$ obtained by defining vertices as centers of the triangles so that faces become vertices and vice versa.

2. The rule $V - E + F = 2(1 - g) - h$, where V, E and F are the numbers of vertices, edges, and faces, relates $V - E - F$ to the topology of the graph, which in the recent case is triangular tesselation. g is the genus of the surface at which the triangulation is im eded and h is the number of holes in it. In case of torus one would have $E = V + F$ giving in the recent case $E = 36$ for $(V, F) = (12, 24)$ (see `http://tinyurl.com/y7xfromc`) whereas in the icosahedral case one has $E = 32$.

3. This kind of tesselations are obtained by applying periodic boundary conditions to triangular lattices in plane defining parallelogram. The intuitive expectation is that this lattices can be labelled by two integers (m, n) characterizing the lengths of the sides of the parallelogram plus angle between two sides: this angle defines the conformal equivalence class of torus. One can also introduce two unit vectors e_1 and e_2 characterizing the conformal equivalence class of torus.

Second naive expectation is that $m \times n \times sin(\theta)$ represents the area of the parallelogram. $sin(\theta)$ equals to the length of the exterior product $|e_1 \times e_2| = sin(\theta)$ representing twice the area of the triangle so that there would be $2m \times n$ triangular faces. The division of the planar lattice by group generated by $pe_1 + qe_2$ defines boundary conditions. Besides this the rotation group Z_6 acts as analog for the symmetries of a unit cell in lattice. This naive expectation need not of course be strictly correct.

4. As noticed, it is not possible to have triangular toric tesselations with $(V, E, F) = (12, 30, 20)$. Torus however has a triangular tesselation with $(V, E, F) = (12, 36, 24)$. An illustration of the tesselation can be found at `http://tinyurl.com/y7xfromc`. It allows to count visually the numbers V, E, F, and the identifications of the boundary edges and vertices. With good visual imagination one might even try to guess what Hamiltonian cycles look like.

   The triangular tesselations and their hexagonal duals are characterized partially by a pair of integers (a, b) and (b, a). a and b must both even or odd (see `http://tinyurl.com/y7xfromc`). The number of faces is $F = (a^2 + 3b^2)/2$. For $(a, b) = (6, 2)$ one indeed has $V = 12$ and $F = 24$. From the article [1] (see `http://tinyurl.com/ya6g9kwe`) one learns that the number of triangles satisfies $F = 2V$ for $p = q$ at least. If $F = 2V$ holds true more generally one has $V = (a^2 + 3b^2)/8$, giving a tight constraints on a and b.

   Remark: The conventions for the labelling of torus tesselation vary. The above convention based on integers (a, b) used in the illustrations at `http://tinyurl.com/y7xfromc` is different from the convention based on integer pair (p, q) used in [1] . In this notation torus tesselation with $(V, F) = (12, 24)$ corresponds to $(p, q) = (2, 2)$ instead of $(a, b) = (6, 2)$. This requires $(a, b) = (3p, q)$. With these conventions one has $V = p^2 + q^2 + pq$.

2.1 The number of triangles in the 12-vertex tesselation is 24: curse or blessing?

One could see as a problem that one has $F = 24 > 20$? Or is this a problem?

1. By fusing two icosahedral harmonies and one toric harmony one would obtain a harmony with 20+20+24 =64 chords, the number of DNA codons! One would replace the fusion of 3 icosahedral harmonies and tetrahedral harmony with a fusion of 2 icosahedral harmonies and toric harmony. Icosahedral symmetry with toric symmetry associated with the third harmony would be replaced with a smaller toric symmetry. Note however that the attachment of tetrahedron to a fixed icosahedral face also breaks icosahedral symmetry.

   This raises questions. Could the presence of the toric harmony somehow relate to the almost exact $U \leftrightarrow C$ and $A \leftrightarrow G$ symmetries of the third letter of codons. This does not of course mean that one could associated the toric harmony with the third letter. Note that in the icosa-tetrahedral model the three harmonies are assumed to have no common chords. Same non-trivial assumption is needed also now in order to obtain 64 codons.

2. What about the number of amino-acids: could it be 24 corresponding ordinary aminoacids, stopping sign plus 3 additional exotic amino-acids. The 20 icosahedral triangles can corresponds to amino-acids but not to stopping sign. Could it be that one of the additional codons in 24 corresponds to stopping sign and two exotic amino-acids Pyl and Sec appearing in biosystems explained by the icosahedral model in terms of a variant of the genetic code. There indeed exists even third exotic amino-acid! N-formylmethionine (see `http://tinyurl.com/jsphvgt`) but is usually regarded as as a form of methionine rather than as a separate proteinogenic amino-acid.

3. Recall that the problem related to the icosa-tetrahedral harmony is that it does not contains the chords of what might be called classical harmonies (the chordds assignable to major and minor

scales). If 24 chords of bio-harmony correspond to toric harmony, one could obtain these chords if the chords in question are chords obtainable by the proposed construction.

But is this construction consistent with the representation of 64 chords by taking to each chord one note from 3 disjoint groups of 4 notes in which each note belongs to 16 chords. The maximum number of chords that note can belong to would be 5+5+6=16 as desired. If there are no common chords between the 3 harmonies the conditions is satisfied. Using for instance 3 toric representations the number would be 6+6+6=18 and would require dropping some chords.

4. The earlier model for tRNA as fusion of two icosahedral codes predicting 20+20=40 tRNA codons. Now tRNAs as fusion of two harmonies allows two basic options depending on whether both harmonies are icosahedral or whether second harmony is toric. These options would give 20+20=40 or 20+24=44 tRNAs. Wikipedia tells that maximum number is 41. Some sources however tell that there are 20-40 different tRNAs in bacterial cells and as many as 50-100 in plant and animal cells.

2.2 A more detailed model for toric harmonies

One can consider also more detailed model for toric harmonies.

1. The above discussed representation in terms of frequencies assigned with nucleotides depending on their position requires the decomposition of the notes to 3 disjoint groups of 4 notes. This means decomposition of 12 vertices of Hamiltonian cycle to 4 disjoint groups such that within given group the distances between the members of group are larger than one unit so that they cannot belong to same triangle. There are $Bin(12,4) \times Bin(8,4)$ decomposition to 3 disjoint groups of for vertices, where $Bn(n,k) = n!/(n-k)!k!$ is binomial coefficient.

2. Once the Hamiltonian cycle has been fixed and is one assumes that single step along cycle corresponds to quint, one knows what the notes associated with each vertex is and given the note of the 12-note scale one knows the number$0 \leq n < 12$ of quint steps needed to obtain it. For instance, for the proposed grouping $\{C, C\sharp, D, D\sharp\}$ and its two transposes by 2 hole steps one can assign 4 integers to each group. The condition is that within each group the notes labelled by the integers have minimum distance of 2 units between themselves.

3. One could try to understand the situation in terms of the symmetries of the system.

   (a) Could the triplet $\{C, E, G\sharp\}$ and its four translates be interpreted as Z_3 orbits. Could suitable chosen members from 4 disjoint quartets quite general form Z_3 orbits.
   Remark: Particle physicists notes the analogy with 4 color triplets formed by u and d quarks having spin 1/2. Z_4 would correspond to spin and color spin and Z_3 to color.

   (b) Z_4 acts as symmetries of the tesselation considered and these symmetries respect distances so that their action on a quartet with members having mutual distances larger than unit creates new such quartet. Could the triplet $\{C, E, G\sharp\}$ and its four translates by an $n-$multiple of half note, $n = 0, 1, 2, 3$ correspond to an orbit Z_4?
   Could the groups of 4 notes quite generally correspond to the orbits of Z_4? This can be true only if the action of non-trivial Z_4 elements relates only vertices with distance larger than one unit.

4. The group of isometries of the toric triangulation acts as symmetries. $Z_{24} = Z_6 \times Z_4$ is a good candidate for this group. Z_6 corresponds to the rotations of around given point of triangulation and should leave the tesselation invariant. The orbit of given triangle defining the set of DNA codons coding the amino-acid represented by the orbit would correspond to orbit of subgroups of Z_{24}. Only orbits containing orbits containing $1, 2, 3, 4$ or 6 triangles are allowed by the degeneracies of the genetic code. These numbers would correspond to degeneracies that is the numbers of codons coding for given amino-acid. All these numbers appear as degeneracies.

2.3 What one can say about toric Hamiltonian cycles?

First some basic notions are in order. The graph is said to be equivelar if it is a triangulation of a surface meaning that it has 6 edges emanating from each vertex and each face has 3 vertices and 3 edges [1]. Equivelarity is equivalent with the folllowing conditions;

1. Every vertex is 6-valent.
2. The edge graph is 6-connected.
3. The graph has vertex transitive automorphism group.
4. The graph can be obtained as a quotient of the universal covering tesselation (3,6) by a sublattice (subgroup of translation group). 6-connectedness means that one can decompose the tesselation into two disconnected pieces by removing 6 or more vertices
5. Edge graph is n-connected if the elimation of $k < n$ vertices leaves it connected. It is known that every 5-connected triangulation of torus is Hamiltonian [2] (see `http://tinyurl.com/y7cartk2`). Therefore also 6-connected $(6,3)_{p=2,q=2}$ tesselation has Hamiltonian cycles.
6. The Hamiltonian cycles for the dual tesselation are not in any sense duals of those for the tesselation. For instance, in the case of dodecahedron there is unique Hamiltonian cycle and for icosahedron has large number of cycles. Also in the case of $(6,3)$ tesselations the duals have different Hamilton cycles. In fact, the problem of constructing the Hamiltonian cycles is NP complete.

Can one say anything about the number of Hamiltonian cycles?

1. For dodecahedron only 3 edges emanates from a given vertex and there is only one Hamiltonian cycle. For icosahedron 5 edges emanate from given vertex and the number of cycles is rather large. Hence the valence and also closely related notion of n-connectedness are essential for the existence of Hamilton's cycles. For instance, for a graph consisting of two connected graphs connected by single edge, there exist no Hamilton's cycles. For toric triangulations one has as many as 6 edges from given vertex and this favors the formation of a large number of Hamiltonian cycles.
2. Curves on torus are labelled by winding numbers (M,N) telling the homology equivalence class of the cycle. M and M can be any integers. Curve winds M (N) times around the circle defining the first (second) equivalence homology equivalence class. Also Hamiltonian cycles are characterized by their homology equivalence class, that is pair (M,N) of integers. Since there are only $V = 12$ points, the numbers (M,N) are finite. By periodic boundary conditions means that the translations by multiples of $2e_1 + 2e_2$ do not affect the tesselation (one can see what this means geometrically from the illustration at `http://tinyurl.com/y7xfromc`). Does this mean that (M,N) belongs to $Z_2 \times Z_2$ so that one would have 4 homologically non-equivalent paths.

   Are all four homology classes realized as Hamiltonian cycles? Does given homology class contain several representatives or only single one in which case one would have 20 non-equivalent Hamiltonian cycles?

It turned out that there exist programs coding for an algorithm for finding whether given graph (much more general than tesselation) has Hamiltonian cycles. Having told to Jebin Larosh about the problem, he sent within five minutes a link to a Java algorithm allowing to show whether a given graph is Hamiltonian (see `http://tinyurl.com/y7y9tr5t`): sincere thanks to Jebin! By a suitable modification this algorithm find all Hamiltonian cycles.

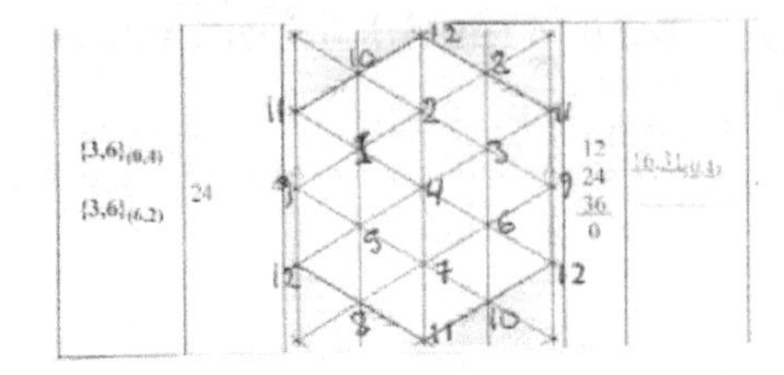

Figure 1: The number of the vertices of $(V, F) = 12, 24)$ torus tesselation allowing path $(0, 1, 2, 3, 4, 6, 5, 8, 10, 7, 11, 9, 0)$ as one particular Hamiltonian cycle.

1. The number N_H of Hamiltonian cycles is expected to be rather large for a torus triangulation with 12 vertices and 24 triangles and it is indeed so: $N_H = 27816$! The image of the tessellation and the numbering of its vertices are described in figure below (see **Fig. 1**). Incide matrix A characterizes the graph: if vertices i and j are connected by edge, one has $A_{ij} = A_{ji} = 1$, otherwise $A_{ij} = Aji = 0$ and is used as data in the algorithm finding the Hamiltonian cycles.

   The cycles related by the isometries of torus tessellation are however equivalent. The guess is that the group of isometries is $G = Z_{2,refl} \rtimes (Z_{4,tr} \rtimes Z_{n,rot})$. $Z_{n,rot}$ is a subgroup of local $Z_{6,rot}$. A priori $n \in \{1, 2, 3, 6\}$ is allowed.

   On basis of [1] I have understood that one has $n = 3$ but that one can express the local action of $Z_{6,rot}$ as the action of the semidirect product $Z_{2,refl} \times Z_{3,rot}$ at a point of tesselation (see `http://tinyurl.com/ya6g9kwe`). The identity of the global actions $Z_{2,refl} \times Z_{3,rot}$ and $Z_{6,rot}$ does not look feasible to me. Therefore $G = Z_{2,refl} \rtimes (Z_{4,tr} \rtimes Z_{3,rot})$ with order $ord(G) = 24$ will be assumed in the following (note that for icosahedral tesselation one has $ord(G) = 120$ so that there is symmetry breaking).

   Z_4 would have as generators the translations e_1 and e_2 defining the conformal equivalence class of torus. The multiples of $2(e_1 + e_2)$ would leave the tesselation invariant. If these arguments are correct, the number of isometry equivalence classes of cycles would satisfy $N_{H,I} \geq N_H/24 = 1159$.

2. The actual number is obtained as sum of cycles characterized by groups $H \subset Z_{12}$ leaving the cycle invariant and one can write $N_{H,I} = \sum_H (ord(H)/ord(G))N_0(H)$, where $N_0(H)$ is the number of cycles invariant under H.

What can one say about the symmetry group H for the cycle?

1. Suppose that the isometry group G leaving the tesselation invariant decomposes into semi-direct product $G = Z_{2,refl} \rtimes (Z_{4,tr} \rtimes Z_{3,rot})$, where $Z_{3,rot}$ leaves invariant the starting point of the cycle. The group H decomposes into a semi-direct product $H = Z_{2,refl} \rtimes (Z_{m,tr} \times Z_{3,rot})$ as subgroup of $G = Z_{2,refl} \rtimes (Z_{4,tr} \times Z_{3,rot})$.

2. $Z_{n,rot}$ associated with the starting point of cycle must leave the cycle invariant at each point. Applied to the starting point, the action of H, if non-trivial - that is $Z_{3,rot}$, must transform the outgoing edge to incoming edge. This is not possible since Z_3 has no idempotent elements so that one can have only $n = 1$. This gives $H = Z_{2,refl} \rtimes (Z_{m,tr}$. $m = 1, 2$ and $m = 4$ are possible.

3. Should one require that the action of H leaves invariant the starting point defining the scale associated with the harmony? If this is the case, then only the group $H = Z_{2,refl}$ would remain and invariance under Z_{refl} would mean invariance under reflection with respect to the axis defined by e_1

or e_2. The orbit of triangle under $Z_{2,refl}$ would consist of 2 triangles always and one would obtain 12 codon doublets instead of 10 as in the case of icosahedral code.

If this argument is correct, the possible symmetry groups H would be Z_0 and $Z_{2,refl}$. For icosahedral code both Z_{rot} and Z_{2_refl} occur but $Z_{2,refl}$ does not occur as a non-trivial factor of H in this case.

The almost exact $U \leftrightarrow C$ and $A \leftrightarrow G$ symmetry of the genetic code would naturally correspond to $Z_{2,refl}$ symmetry. Therefore the predictions need not change from those of the icosahedral model except that the 4 additional codons emerge more naturally. The predictions would be also essentially unique.

4. If H is trivial Z_1, the cycle would have no symmetries and the orbits of triangles would contain only one triangle and the correspondence between DNA codons and amino-acids would be one-to-one. One would speak of disharmony. Icosahedral Hamiltonian cycles can also be of this kind. If they are realized in the genetic code, the almost exact $U \leftrightarrow C$ and $A \leftrightarrow G$ symmetry is lost and the degeneracies of codons assignable to 20+20 icosahedral codons increase by one unit so that one obtains for instance degeneracy 7 instead of 6 not realized in Nature.

What can one say about the character of toric harmonies on basis of this picture.

1. It has been already found that the proposal involving three disjoint quartets of subsequent notes can reproduce the basic chords of basic major and minor harmonies. The challenge is to prove that it can be assigned to some Hamiltonian cycle(s). The proposal is that the quartets are obtained by Z_{rot}^3 symmetry from each other and that the notes of each quartet are obtained by $Z_{4,tr}$ symmetry.

2. A key observation is that classical harmonies involve chords containing 1 quint but not 2 or no quints at all. The number of chords in torus harmonies is $24 = 2 \times 12$ and twice the number of notes. The number of intervals in turn is 36, 3 times the number of the notes. This allows a situation in which each triangle contains one edge of the Hamiltonian cycle so that all 3-chords indeed have exactly one quint.

3. By the above argument harmony possesses Z_2 symmetry or no symmetry at all and one has 12 codon doublets. For these harmonies each edge of cycle is shared by two neighboring triangles containing the same quint. A possible identification is as major and minor chords with same quint. The changing of the direction of the scale and the reflection with respect to the edges the Hamiltonian cycle would transforms major chords and minor chords along it to each other and change the mood from glad to sad and vice versa.

   The proposed harmony indeed contains classical chords with one quint per chord and for $F, A, C\sharp$ both minor and major chords are possible. There are 4 transposes of this harmony.

4. Also Hamiltonian cycles for which n triangles contain two edges of Hamiltonian path (CGD type chords) and n triangles contain no edges. This situation is less symmetric and could correspond to a situtation without any symmetry at all.

5. One can ask whether the classical harmonies corresponds to 24 codons assignable to the toric harmony and to the 24 amino-acids being thus realizable using only amino-acids. If so, the two icosahedral harmonies would represent kind of non-classical exotics.

3 Appendix: Some facts about toric tesselations

Genus $g = 1$ (torus) is unique in that it allows infinite number of tesselations as analogs of planar lattices with periodicic boundary conditions. $g = 0$ allows only Platonic solids as tesselations and $g > 1$ allows very few tesselations. The article [1] gives a nice review about toric tesselations.

1. Toric tesselations correspond to tesselations of plane by periodic boundary conditions. Torus tesselation allows a universal covering identifiable as counterpart of infinite lattice in plane. There are infinite number of coverings of given tesselation labelled by two integers (m, n) since the homology group of torus is $Z \times Z$. The tesselation is obtained by dividing $Z \times Z$ by its normal subgroup. Also the rotation group Z_6 acts as group leaving the tesselation invariant and correspond to the rotation leaving invariant the lattice cell consisting of 6 vertices around given vertex.

2. The tesselation is called decomposable if there is a k-sheeted covering map (map corresponds to a collection of charts) characterized by the subgroup of the isometries of the covering of the tesselation which corresponds to a sub-tesselation. This subgroup is charactrized by a pair (p, q) of integers being generated by the translation $pe_1 + qe_2$ and $2\pi/6$ rotation. The unit vectors can be chosen to be $e_1 = (1, 0)$ and $e_2 = (1, \sqrt{3})/2$ for triangular tesselation (presumably this tesselation is regular tesselation with the conformal equivalence class of torus fixed by the angle between e_1 and e_2). Line reflection transforms $(3, 6)_{p,q}$ to $(3, 6)_{q,p}$ (see Fig 1 of `http://tinyurl.com/ya6g9kwe`). The tesselation is invariant under reflections - regular -if $pq(p-q) = 0$. The peculiar looking form of the conditions follows from the identitity $(3, 6)_{q,p} = (3, 6)_{p+q,-q}$ (also $p = 0$ or $q = 0$ is possble) Note that the tesselation $(3, 6)_{2,2}$ is invariant under reflection and thus non-chiral.

3. The number V of vertices of the triangular itesselation is given by $V = p^2 + q^2 + pq$. The regular tesselation $(p, q) = (2, 2)$ has 12 vertices and is the interesting one in the recent case. It is the smallest regular tesselation. For given (p, q) one can have several non-equivalent pairs (p, q) defining combinatorially non-equivalent tesselations. My interpretation is that they correspond to different conformal equivalence classes for torus: the intuitive expectation is that this should not affect the topology of tesselation nor Hamiltonian cycles. For $(6, 3)_{p,q} = (6, 3)_{2,2}$ with s $(V = 12, F = 24)$ there are 1+6 =7 combinatorially non-equivalent tesselations: one non-chiral and 6 chiral ones.

   Quite generally, the tesselations with V vertices with $V \ mod \ 4 = 0$ (as in the case of $V = 12$) allow one map (chart consisting of faces) with isotropy group of order 2 and 6 maps with isotropy group of order 4. These variants are labelled by an SL(2,Z) matrix $(a, b; 0, c)$ with determinant equal to $V = ac$. For $V = 12$ one has decompositions $12 = 1 \times 12$, $12 = 2 \times 6$, $12 = 3 \times 4$. $-c < b < a - c$ is unique modulo a. In the recent case one as $ac = 12$ allowing $(a, c) \in \{(1, 12), (2, 6), (3, 4)\}$ and pairs obtained by permuting a and c. These matrices need not define combinatorially different tesselations since modular transformations generate equivalent matrices.

Acknowledgements: I am grateful to Jebin Larosh for finding the Java algorith allowing to find Hamilton cycles for given graph.

References

[1] Ulrich Brehm and Wolfgang Kühnel. Equivelar maps on the torus. *European Journal of Combinatorics*. *https://doi.org/10.1016/j.ejc.2008.01.010*, 29(8):1843–1861, 2008.

[2] Thomas R and Yu X. Five-Connected Toroidal Graphs Are Hamiltonian. *Journal of Combinatorial Theory, Series B*, 69(TB961713):79–96, 1997.

[3] Pitkänen M. Geometric theory of harmony. Available at: `http://tgdtheory.fi/public_html/articles/harmonytheory.pdf`, 2014.

[4] Pitkänen M. New results in the model of bio-harmony. Available at: `http://tgdtheory.fi/public_html/articles/harmonynew.pdf`, 2018.

Exploration

Could Cancer Be a Disease of Magnetic Body?

Dana Flavin [1] and Matti Pitkänen [2]

Abstract

Li and Heroux have made a highly interesting discovery. The treatment of cancer cell population by 60 Hz oscillating magnetic field with extremely small strength above 25 nT leads to a reduction of the abnormally large chromosome number of the mitochondria of cancer cells and eventually the cancer cells return to the normal state. TGD based explanation for the findings relies on the basic notions of TGD inspired quantum biology. The basic notions are magnetic body (MB) and hierarchy of Planck constants $h_{eff} = n \times h_0$ ($h = 6h_0$) emerging from the adelic physics as a prediction but originally proposed on basis of anomalous effects of ELF em fields in living matter. The value of n can be relatively small or very large corresponding to flux tubes mediating em and gravitational interactions. The anatomy of MB has remained unclear hitherto but in this article a detailed model is developed allowing to understand the formula $h_{gr} = h_{eff} = n_{gr}h_0$ for gravitational Planck constant and leading to a further formula for h_{gr} relating magnetism and gravitation. A further central notion is TGD based model for water memory as the ability of the MB of water to control the thickness of its flux tubes to entrain with external frequencies and reproduce them. This is a central element in the TGD based view about immune systems and homeopathic effects. Cancer would reduce to a disease of the MB of the system, to a high degree determined by MB of water and homeopathy like treatment based on irradiated water could serve as a cure. The model is applied both to the findings of Montagnier's group about remote regeneration of DNA without template and to those of Li and Heroux.

Keywords: Cancer, disease, magnetic body, quantum biology, TGD framework.

1 Introduction

I received from Prof. Dana Flavin a link to an extremely interesting popular article "*A Unified Theory of Weak Magnetic Field Action*" by L Slesin, editor of Microwave News published in in EMFact (see `http://tinyurl.com/yd4jpuq6`). The article tells about the work of karyologists Ying Li and Paul Heroux. Karyology is a branch of biology studying chromosomes. The research article "*Extra-low-frequency magnetic fields alter cancer cells through metabolic restriction*" by Li and Heroux is published "*Electromagnetic Biology ad Medicine*" [8] (see `http://tinyurl.com/y9lv47qp`).

1.1 Experimental findings of Li and Heroux

The abstract of the article is following.

"**Background**: *Biological effects of extra-low-frequency (ELF) magnetic fields (MFs) have lacked a credible mechanism of interaction between MFs and living material. Objectives: To examine the effect of ELF-MFs on cancer cells. Methods: Five cancer cell lines were exposed to ELF-MFs within the range of 0.0255 μT, and the cells were examined for karyotype changes after 6d.*

Results: *All cancer cells lines lost chromosomes from MF exposure, with a mostly flat dose-response. Constant MF exposures for three weeks allow a rising return to the baseline, unperturbed karyotypes. From this point, small MF increases or decreases are again capable of inducing karyotype contractions (KCs). Our data suggest that the KCs are caused by MF interference with mitochondrias adenosine*

[1] Correspondence: Dana Flavin, Eschenweg 4, 82319 Sternberg, Germany. Email: DANA_FK@hotmail.com.
[2] Matti Pitkänen `http://tgdtheory.com/`. Address: Rinnekatu 2-4 8A, 03620, Karkkila, Finland. Email: matpitka6@gmail.com.

triphosphate synthase (ATPS), compensated by the action of adenosine monophosphate-activated protein kinase (AMPK). The effects of MFs are similar to those of the ATPS inhibitor, oligomycin. They are amplified by metformin, an AMPK stimulator, and attenuated by resistin, an AMPK inhibitor. Over environmental MFs, KCs of various cancer cell lines show exceptionally wide and flat dose-responses, except for those of erythroleukemia cells, which display a progressive rise from 0.025 to 0.4T.

Conclusions*: The biological effects of MFs are connected to an alteration in the structure of water that impedes the flux of protons in ATPS channels. These results may be environmentally important, in view of the central roles played in human physiology by ATPS and AMPK, particularly in their links to diabetes, cancer and longevity.*"

Li and Heroux count the number of chromosomes in cancer cell population before and after a irradiation with 60 Hz oscillating magnetic field, which is extremely weak, with strength above 25 nT. Unlike normal human cells with 46 chromosomes, cancer cells have a variable number of chromosomes (typically this causes trouble, Down's syndrome is one example). Plants have very often this kind of replication of chromosomes and have isolated cells. Could the replication be due to an effective isolation of cancer cells from each other due caused by a loss of coherent behavior.

Cancer cells typically 74 chromosomes. Li and Heroux report that the irradiation using extremely weak magnetic 60 Hz fields as low as 25-50 nT for 6 days, the cells lose 190 per cent of their chromosomes. They call the effect karyotype contraction.

They repeated the experiment with 4 other cell lines - lung and colon cancer and two different types of leukemias and found essentially the same effect every time.

1. After 3 weeks on the field, the number of chromosomes returns to baseline numbers.

2. Once adapted to the magnetic field, the cells become exquisitely sensitive to further variations of the magnetic field. An increase or decrease of only 10 nT will prompt another round of karyotype contractions.

3. The karyotype contractions vary very little over a wide range of field intensities: from 100 500 nT so that only frequency seems to matter.

1.2 The proposed interpretation of findings of Li and Heroux

The interpretation of the findings in the standard physics framework is far from obvious. So weak magnetic fields should not have so strong effects on cancer cell population. Li and Heroux locate the problem to ATPase molecules acting as kind of power plants of cell associated with mitochondrial membrane. ATPase pumps protons against potential gradient and when protons return back, they provide their energy for the formation of of ATP serving as the metabolic energy currency of cell. The basic observation is that that the impairing the action of ATPase leads also to karyotype contraction and in the long run makes cells normal. This happens if the cell does not receive enough oxygen in which case anaerobic metabolism starts and produces lactic acid.

From this Li and Heroux conclude that the problem is too high rate for the metabolic energy and that the irradiation reduces the metabolic energy feed by somehow changing the properties of ATPase. They suggests that the weak oscillating magnetic field affects the physical properties of water in the ATPase. The flow of protons becomes slower and the rate of metabolism becomes slower. One could imagine that the conductivity of protons is reduced.

Li and Heroux refer to the article (see `http://tinyurl.com/yd7kqnzg`) *Effect of weak magnetic fields on the properties of water and ice* by Russian physicists L Semikina and V Kiselev [3] (see `http://tinyurl.com/yd7kqnzg`). They show that magnetic fields with strengths in the range 7 Gauss - 2 nT and with frequencies in the range .01-200 Hz have measurable effects on the properties of water. A possible mechanism would be reduction of proton conductivity.

The abstract of the article is here.

"We establish that a number of physical properties of water and ice are significantly changed by an alternating magnetic field of a certain frequency. The changes in the physical parameters of ice are several times stronger than the changes in the corresponding parameters of water. Heating water to 50 ° C destroys the magnetic effects. When the field is much weaker than the geomagnetic field, a change in water purity (bidistilled instead of distilled water) only broadens the extrema observed in the state dependences of water and ice on the frequency of the alternating magnetic field of constant amplitude. The magnitude and intensity of these extrema are unaffected by water purity. The effects of the geomagnetic field on the properties of ice and water are also discussed."

One important observation is that the effects of ELF em fields depend very weakly on strength but depend on frequency. In standard physics it is difficult to understand this.

1.3 Criticism

There are at least two objection against the interpretation of the findings proposed by Li and Heroux.

1. First objection relates to what happens in real cancer as compared to cancer in cell lines. One can argue that cell lines is different from real cancer tissue because they live in a "luxury" whereas real cancer cells suffer from the lack of oxygen. Reduction of metabolic energy feed seems to bring the cells to normal state. Could too high feed of metabolic energy transform ordinary cells to cancer cells? Why? Could the primary reason for cancer be something else than too high metabolic energy feed and be in action also in the real cancer tissue suffering from too low oxygen feed. Could the too low oxygen feed be an attempt of organism to get rid of cancer cells or force them to act like ordinary cells? What could this primary reason for cancer be?

2. Physicist inside me objects strongly the proposed explanation for the return of normal state. Reduction of metabolism could explain it but the magnetic fields seem quite too weak to reduce the rate of proton flow.

A leading example is provided by sociology - I have a concrete experience from my own country at 90's. Things were extremely well but then something strange happened. People became extremely greedy and selfish and kind of manic consumption emerged. Intensive financial speculation began and eventually led to a very bad economical depression. Trust disappeared from the society and social structures started to decay, and loneliness is the basic problem of quite many people nowadays. Social coherence was lost and during decades the situation has become even worse as society has split into rich and poor. Civilizations have life cycle beginning with healthy social structures and ending up to a deeply corrupted deeply divided civilization eventually collapsing. The collapse of civilization means very hard times for individuals but leads to a new civilization with healthy social structures. Do good times have the effect that people do not need each other anymore and this leads to a loss of coherence.

What happened was very much like cancer in which tissue decomposes into cells with only one goal which is to replicate. In the case of bacteria populations it has been found that starvation leads to a formation of analogs of multi-cellulars. The opposite happens in the case of cancer.

In a framework of quantum physics based theory of consciousness predicting a hierarchy of conscious entities [25] it is rather natural to apply same principles in attempts to understand the behaviour of human society and cell community. Could hard times have a healthy effect also on cancer cell line living in luxury? Could the primary reason for the return of normal state be the generation of coherent less selfish behavior of cells leading also to a restriction of metabolism.

What the coherent behaviour of the population does mean physically, what does induce it? Standard physics does not provide any obvious answer.

1.4 TGD based model

TGD based model for the findings relies on some central aspects of TGD inspired quantum biology summarized first. Some new important details related to the magnetic body (MB), dark matter hierarchy labelled by the value of Planck constant h_{eff}, and the notion of gravitational Planck constant h_{gr} are discussed. The TGD based view about water memory and homeopathy are introduced. The basic mechanism is the entrainment of the MB of water to frequencies of external em signals by varying the thickness of its flux tubes so that cyclotron frequencies are tuned to resonance. Kind of living radio set also able to serve as a sender is in question.

The model is applied to two situations. The first application is to the earlier findings by Montagnier's group [7] about remote regeneration of DNA discussed already earlier in TGD context [36]. There are two samples A and B. The remote regeneration of DNA occurs in B in absence of template and without any physical contact to A serving effectively as a template. A contains originally the DNA but is extremely diluted. Second application is a model of cancer applied to the experiments of Li and Heroux. The MB of water would go out of synch from central control frequencies or lose part of its MB generating these frequencies. The proposed healing mechanism would be "homeopathic" treatment by water entrained to the missing control frequencies.

2 Some aspects of TGD inspired quantum biology

TGD based explanation for the findings relies on the basic notions of TGD inspired quantum biology. The basic notions are magnetic body (MB) and hierarchy of Planck constants $h_{eff} = n \times h_0$ [30, 31] emerging from the adelic physics as a prediction [48, 49] but originally proposed on basis of anomalous effects of ELF em fields in living matter. The anatomy of MB has remained unclear hitherto but in this article a detailed model allowing to understand the formula $h_{gr} = h_{eff}$ for gravitational Planck constant and leading to a further formula for h_{gr} relating magnetism and gravitation.

A further central notion is TGD based model for water memory as the ability of the MB of water to control the thickness of its flux tubes to entrain with external frequencies and reproduce them. This is a central element in TGD based view about immune system and homeopathic effects [17]. Cancer would reduce to a disease of the MB of the living system to high degree determined by the MB of water. Details of the bio-chemistry and even cell membrane dynamics would have surprisingly minor role in the model.

2.1 The notion of magnetic body

Magnetic flux tubes and field body/magnetic body (MB) are basic notions of TGD implied by the modification of Maxwellian electrodynamics [27, 18, 24]. Actually a profound generalization of space-time concept is in question. Magnetic flux tubes are in well-defined sense building bricks of space-time - topological field quanta - and lead to the notion of field body/MB as a field identity assignable to any physical system: in Maxwell's theory and ordinary field theory the fields of different systems superpose and one cannot say about magnetic field in given region of space-time that it would belong to some particular system. In TGD only the effects on test particle for induced fields associated with different space-time sheets with overlapping M^4 projections sum.

The hierarchy of Planck constants $h_{eff} = n \times h_0$, where h_0 is the minimum value of Planck constant, is second key notion. h_0 need not correspond to ordinary Planck constant h and both the observations of Randell Mills [42] and the model for color vision [54] suggest that one has $h = 6h_0$. The hierarchy of Planck constants labels a hierarchy of phases of ordinary matter behaving as dark matter.

Magnetic flux tubes would connect molecules, cells and even larger units, which would serve as nodes in (tensor-) networks [1][41]. Flux tubes would serve as correlates for quantum entanglement and replace wormholes in ER-EPR correspondence proposed by Leonard Susskind and Juan Maldacena in 2014 (see `http://tinyurl.com/y7za98cn` and `http://tinyurl.com/ydckw5u7`). In biology and neuroscience these networks would be in a central role. For instance, in brain neuron nets would be associated with them and would serve as correlates for mental images [45, 55]. The dynamics of mental images would correspond to that for the flux tube networks.

2.2 Hierarchy of Planck constants, space-time surfaces as covering spaces, and adelic physics

From the beginning it was clear that $h_{eff}/h = n$ corresponds to the number of sheets for a covering space of some kind. First the covering was assigned with the causal diamonds. Later I assigned it with space-time surfaces but the details of the covering remained unclear. The final identification emerged only in the beginning of 2017.

2.2.1 Number theoretical universality and hierarchy of extensions of rationals

Number theoretical universality (NTU) leads to the notion of adelic space-time surface (monadic manifold) involving a discretization in an extension of rationals defining particular level in the hierarchy of adeles defining evolutionary hierarchy. The formulation of this vision is proposed in [44, 49, 48].

The key constraint is NTU for adelic space-time containing sheets in the real sector and various padic sectors, which are extensions of p-adic number fields induced by an extension of rationals which can contain also powers of a root of e inducing finite-D extension of p-adic numbers (e^p is ordinary p-adic number in Q_p).

One identifies the numbers in the extension of rationals as common for all number fields and demands that imbedding space has a discretization in an extension of rationals in the sense that the preferred coordinates of imbedding space implied by isometries belong to extension of rationals for the points of number theoretic discretization. This implies that the versions of isometries with group parameters in the extension of rationals act as discrete versions of symmetries. The correspondence between real and p-adic variants of the imbedding space is extremely discontinuous for given adelic imbedding space (there is hierarchy of them with levels characterized by extensions of rationals). Space-time surfaces typically contain rather small set of points in the extension ($x^n + yn^2 = z^n$ contains no rationals for $n > 2!$). Hence one expects a discretization with a finite cutoff length at space-time level for sufficiently low space-time dimension $D = 4$ could be enough.

After that one assigns in the real sector an open set to each point of discretization and these open sets define a manifold covering. In p-adic sector one can assign 8:th Cartesian power of ordinary p-adic numbers to each point of number theoretic discretization. This gives both discretization and smooth local manifold structure. What is important is that Galois group of the extension acts on these discretizations and one obtains from a given discretization a covering space with the number of sheets equal to a factor of the order of Galois group.

2.2.2 Effective Planck constant as dimension of extension of rationals and number of sheets of space-time surface as covering space

$h_{eff}/h_0 = n$ was identified from the beginning as the number of sheets of poly-sheeted covering assignable to space-time surface. The number n of sheets would naturally a factor of the order of Galois group implying $h_{eff}/h = n$ bound to increase during number theoretic evolution so that the algebraic complexity increases. Note that WCW decomposes into sectors corresponding to the extensions of rationals and the dimension of the extension is bound to increase in the long run by localizations to various sectors in self measurements [20]. Dark matter hierarchy represents number theoretical/adelic physics and therefore has now rather rigorous mathematical justification. It is however good to recall that $h_{eff}/h = n$ hypothesis emerged from an experimental anomaly: radiation at ELF frequencies had quantal effects of vertebrate brain impossible in standard quantum theory since the energies $E = hf$ of photons are ridiculously small as compared to thermal energy.

Indeed, since n is positive integer evolution is analogous to a diffusion in half-line and n unavoidably increases in the long run just as the particle diffuses farther away from origin (by looking what gradually happens near paper basket one understands what this means). The increase of n implies the increase of maximal negentropy and thus of negentropy. Negentropy Maximization Principle (NMP) follows from adelic physics alone and there is no need to postulate it separately. Things get better in the long run although we do not live in the best possible world as Leibniz who first proposed the notion of monad proposed!

2.2.3 Formula for the gravitational Planck constant and some background

The formula

$$\hbar_{gr} = \frac{GM_D m}{v_0} \qquad (2.1)$$

for the gravitational Planck constant was originally introduced by Nottale [2].In m mass of planet whereas M is large mass Here v_0 is a parameter with dimensions of velocity: I have considered argument allowing to deduce information about the value of $\beta_0 = v_0/c$ as the ratio of the M^4 size of the system and the size of its magnetic body [51]. Values of order $\beta_0 \sim 10^{-3}$ are encountered.

Since m disappears from the predictions by Equivalence Principle it is not at all clear what kind limitations one has for m and one can even assume that m corresponds to particle mass without change in predictions. In Nottale's original formula m is mass of planet and M_D the mass of Sun but m could be even mass of elementary particle without change in predictions. The assumption has been $m/M_D << 1$. The replacement of M_D with total mass $M_D + m$ and m by reduced mass $M_D m/(M_D + m)$ does not affect the formula and the asymmetry between m and M_D would become more natural asymmetry between total mass and reduced mass.

For $Mm < v_0 m_{Pl}^2$ one must have $h_{gr} = h$, which suggests that quite generally one must have $m \geq \sqrt{v_0} M_{Pl}$ and $M \geq \sqrt{v_0} M_{Pl}$. The formula is non-relativistic but one can consider a relativistic generalization in which m and M are replaced by energies [34] .

The formula is expected to hold true at the magnetic flux tubes mediating gravitational interaction. M_D has been interpreted as dark gravitational flux at the gravitational flux tubes with a fixed value of h_{eff} and should be a fraction of the total gravitational flux M. These flux tubes define $n_{gr} = h_{eff}/h_0$-sheeted covering of M^4.

Also a more general formula

$$h_{gr} = h_{eff} \ , \quad h_{eff} = n_{gr} \times h_0 \ , \quad h = 6h_0 \ . \qquad (2.2)$$

has been assumed. The support for the formula $h = 6h_0$ is discussed in [42, 54]. The value of h_{gr} can be very large unlike the alue of h_{eff} associated with say valence bonds.

One important implication of the formula is that the cyclotron energy spectrum does not depend on the mass of charged particle at all and is therefore universal. The assumption has been that the spectrum is in visible and UV range assignable to bio-photons [28, 29]. One can however consider also the possibility that also the energies between the thermal energy at physiological temperature and visible photon energies are allowed.

2.2.4 New constraint between h_{gr} and h_{eff}

Cyclotron frequencies and energies in magnetic field B and charged particle with charge Ze and mass m are proportional to the ZeB/m. The energy spectrum of bio-photons would be covered by a spectrum of magnetic field strengths B. A special field strength $B_{end} = 0.2$ Gauss has emerged in biological applications from the beginning and the first guess is that it defines a lower bound for the spectrum of visible photon energies [53, 50, 58]. One can fix the value of h_{gr} and therefore of GM_D/v_0 if one requires that dark photon frequency of say $f_l = 10$ Hz corresponds to the lower bound $f_h = 400$ THz for visible frequencies as $h_{gr} = f_h/f_l$: in this case would would have $n_{gr} = 4 \times 10^{13}$.

The variation of B means variation of cyclotron frequency and I have proposed that the audible frequencies correspond to a spectrum of B for the flux tubes involved with hearing [22], and that even 12-note scale represent in terms of rational frequency ratios might have a preferred role [37, 57].

The formula $h_{gr} = h_{eff}$ is not enough to fix the model completely. A formula fixing the relationship between B and GM_D/v_0 would be needed. This formula should be consistent with $h_{gr} = h_{eff}$. Dimensional analyst would start from the geometry of the situation.

Magnetic flux tubes are characterized by two parameters: length L_c and radius R_B.

1. Length scale naturally corresponds to the cyclotron wave length

$$L_c = \lambda_c = \frac{1}{f_c} = \frac{2\pi m}{ZeB} \ . \tag{2.3}$$

   L_c is proportional to the mass m of the charged particle so that charge particles with different mass are with different mass flux tubes with different length and therefore different onion-like layers of MB. Charged dark particles are like books about different topics at different shelves so that living matter is extremely well-organized: something totally different from a chaotic soup of charged ions.

2. The radius of the flux tube is obtained from the flux quantization. For ordinary cylindrical flux tube with constant B the condition is $BS = k\hbar$ and for $S = \pi R^2$ the radius would be

$$R_B(h,k) = \sqrt{\frac{k\hbar}{\pi eB}} = \sqrt{\frac{k}{\pi}} L_B \ , \quad L_B = \sqrt{\frac{\hbar}{eB}} \ . \tag{2.4}$$

   .

   For $k = 1$ and for $B = B_{end} = .2$ Gauss one has $R_B(h,1) = 3.3$ μm to be compared with p-adic length scale $L(167) = 2.5$ μm assignable to Gaussian Mersenne $M_{G,167} = (1+i)^{167} - 1$. Magnetic length L_B is in this case $L_B = 5.8$ μm slightly larger than $L(169)$.

3. For $h_{eff} = n \times h_0$, $h = 6h_0$ the formula would generalize to

$$R_B(h_{eff},k) = \sqrt{\frac{k\hbar_{eff}}{\pi eB}} = \sqrt{\frac{n}{6}} R_c(h,k) = \sqrt{\frac{nk}{6}} R_B(h,1) \ . \tag{2.5}$$

   Note that here n is rather small such as the value of n assignable to valence bonds.

4. The natural guess is that this formula applies at the small part of the MB restricted to the "biological body" of the living system defining that part of system, which corresponds to relatively small values of h_{eff}. The value of h_{eff} would indeed vary, being larger than h for instance for valence bonds [46]. For dark flux tubes with small value of n the radius would be scaled up by $\sqrt{n}$ such as biological system for fixed value of B. Same happens if the value of flux is scaled by m.

   For the simplest flux tubes carrying monopole flux having string world sheet as M^4 projection geodesic sphere as CP_2 projection, the cross section is not circular disk but CP_2 geodesic sphere with radius R. I this case R is fixed. The M^4 projection of these objects is however unstable against thickening and for spherical cross section- think of two disks glued along boundaries but having different CP_2 projections, the area is $4\pi R^2$, where R corresponds to the radius of M^4 projection. Area is reduced by factor 4 from that for non-monopole flux tube and radius is reduced by factor 1/2.

One can guess the additional constraint on h_{gr} without more detailed analysis of what MB really is using dimensional analysis and I will postpone this analysis later.

1. The first natural guess is that one has

$$\frac{h_{gr}}{h_0} = n_{gr} = x\frac{L_c}{R_B(h_{eff},k)} = x(6\pi)^{3/2}\frac{1}{(nk)^{1/2}}\frac{L_B}{l_C(m)} ,$$
$$L_B = \sqrt{\frac{\hbar}{eB}} , \quad l_C(m) = \frac{\hbar}{m} . \tag{2.6}$$

   x is some numerical constant. h_{gr}/h_0 is proportional to the ratio l_B/l_C of the magnetic length and Compton length $l_C = m/\hbar$ of the charged particle.

2. Alternative guess replaces the radius of the magnetic flux tube with the magnetic length L_B.

$$\frac{h_{gr}}{h_0} = n_{gr} = x\frac{L_c}{L_B} == x6^{3/2}\pi\frac{1}{n^{1/2}}\frac{L_B}{L_C(m)} , \tag{2.7}$$

   This formulas is related by factor $\sqrt{kpi}$ the first formula and has no dependence on h. It is difficult to say anything about exact value of the numerical constant x.

3. h_{gr} is proportional to m so that the formulas are consistent with $h_{gr} = h_{eff}$ formula. Combining these formulas one obtains

$$\frac{GM_D}{h_0 v_0} = \frac{r_S(M_D)}{2} = x2\pi\sqrt{\frac{n}{6Z}}\sqrt{\frac{\hbar}{eB}} . \tag{2.8}$$

   This formula does not depend on m and gives the value of GM_D/v_0 assignable to the flux tubes carrying magnetic field with strength B and particles with charge Z. One can say that the Schwartschild radius $r_S = 2GM_D$ characterixing M_D is proportional to magnetic length. The first option gives

$$r_S(M_D) = x \times 2 \times 6^{1/2}\pi^{3/2}\frac{1}{(nk)^{1/2}}v_0 l_B . \tag{2.9}$$

   For Earth Schwartschild radius is $r_{S,E} = 8.87$ mm and if $M_D < M_E$ holds true, one obtains for a given value of v_0 upper bound for the magnetic length and therefore lower bound for B. I have considered in [51] a model for v_0 and combining this model for this formula, one obtains rather strong constraints on the parameters and also on the minimal value of B. The order of magnitude for v_0 is $v_0 \sim 10^{-3}$.

   M_D/v_0 would not depend on the mass of the charged particles at the flux tube (universality) but would depend on their charge Z unless the parameter x has a compensating $Z-$dependence. Therefore electrons and their Cooper pairs would have different value of GMD/v_0. One could perhaps interpret r_S/v_0 as analog of star radius applying to particular dark matter part of Earth. It would be considerably larger than Schwartchild radius.

4. Note that the condition $GM_D m/v_0 = n_{gr}\hbar$ can be written as

$$r_S(M_D) = 2n_{gr}l_C . \tag{2.10}$$

2.2.5 CP_2 length scale as Planck length and dark mass M_D

The estimates for M_D based on $n_{gr} = h_{gr}/h = G_N M_D m/\hbar v_0$ give for $n_{gr} = 10^{13}$ the estimate in the case of proton $M_D = n_{gr}\hbar v_0/G_N m_p = 3.6 \times v_0 \times M_E$ - a considerable fraction of the mass of Earth!

An additional condition to h_{gr} would come from the proposal [52] that Planck length l_P is actually CP_2 radius, that is radius of CP_2 geodesics sphere (apart from numerical constant) and Newton's constant G is given as

$$G = \frac{l_P^2}{\hbar_{gr}} = \frac{R^2}{\hbar_{gr}} \quad . \tag{2.11}$$

This proposal is extremely attractive, even compelling in TGD framework and would require that gravitational Planck constant has varying value. If the size of the orbits scales like h_{gr} and Newton's constant G scales like $1/h_{gr}$ one has scaling invariance of Newton's equations. The standard value of Newton constant would correspond to $h_{gr}/h \equiv x_N \sim 10^7$ for $G = G_N$. The proposal could explain the variation of Newton's constant and fountain effect in super-conductivity would be one application [30].

For these reasons it is highly interesting make an estimate for M_D assuming that G is dynamical and proportional to $1/h_{gr}$ to see whether the proposals are consistent with each other.

By combining the conditions $G = R^2/\hbar_{gr}$ and $\hbar_{gr} = GM_D m/v_0$, one would obtain the condition

$$\hbar_{gr} = \sqrt{\frac{R^2 M_D m}{v_0}} \quad . \tag{2.12}$$

It however turns out that the latter formula leads to non-realistic estimate for M_D for particle masses such as proton mass $m = m_p$. The reason is that small value of m for given value of M_D implies larger value of M_D. In the original formula of Nottale m was taken as planet mass and M as solar mass. One could argue that the dark gravitational flux tubes connected dark masses and these consist of relatively large number of ordinary particles forming a quantum coherent unit.

To get more perspective it is good to perform a concrete estimate. A more concrete formula is obtained by parameterizing particle mass m in terms of its p-adic length scale: $m = x_m p^{-1/2} = x_m 2^{-k(m)/2}\hbar/R$ (p-adic length scale hypothesis $p \simeq 2^k$ has been used). Proton corresponds to Mersenne prime $p = M_{107} = 2^{107} - 1$. It is convenient to parametrize m as a multiple of proton mass $m = y(m)m_p$.

1. Dark mass for which we do not have detailed identification, can be parameterized as

$$M_D = N(A)Am_p$$

by approximating matter as condensed matter with atom mass $M = Am_p$, A the mass number. This would give

$$n_{gr} = \frac{\hbar_{gr}}{\hbar_0} = 6\sqrt{\frac{y(m)x(m)x(p)^2 x_N A}{v_0}}\sqrt{N(A)} \times 2^{(-k(m)-107)/4} \quad . \tag{2.13}$$

2. The condition $n_{gr}/6 > 1$ (unless this condition is true, one has $h_{gr} = h = 6h_0$) gives

$$N(A) \geq \frac{v_0}{36y(m)x(p)^2 x_N A} \times 2^{(k(m)+107)/2} = \frac{v_0}{36y(m)x(p)^2 x_N A} \times 2^{107} 2^{(k(m)-107)/2} \quad . \tag{2.14}$$

One has $x(p) = 4.2$ from proton-electron mass ratio and p-adic mass formula for electron having $p = M_{127}$.

3. For proton the condition $n_{gr}/6 \geq 1$ would give

$$N(A) \geq \frac{v_0}{36y(m)x(p)^2 x_N A} \times 2^{107} \sim \frac{v_0}{36y(m)x(p)^2 x_N A} \times 1.6 \times 10^{32} \ . \tag{2.15}$$

For water with density $\rho = 10^3$ kg/m^3 giving 1.67 proton masses per Angstrom^3, one would have $N(A) = \rho V/Am_p \equiv \rho d^3/Am_p$. This would give

$$d \geq (\frac{m_p}{\rho})^{1/3}(\frac{v_0}{36y(m)x(p)^2 x_N})^{1/3} \times (1.6 \times 10^{32})^{1/3} \ .$$

For $v_0 \sim 10^{-3}$ and $y(m) = 1$ this would give for $n_{gr} \geq 1$ $d \geq 33$ Mm, which is a considerably larger than Earth radius $R_E = 6.4\times$ Mm.

The mathematical reason for the failure is that $m = m_p$ is too small so that given value of GM_Dm requires a large value M_D. In the original formula of Nottale m was taken as planet mass and M as solar mass. One could argue that the dark gravitational flux tubes connected dark masses and these consist of relatively large number of ordinary particles forming a quantum coherent unit.

1. The lower bound for m as ordinary Planck mass $m_P \sim 10^{18} m_P$ is what comes first in mind since one can argue that $m > m_P$ $h_{gr} > h$ is possible. This would reduce the estimate for d by factor 10^{-6} to $d \sim 33$ m.

2. In the proposed framework a more plausible option comes from the identification of $l_P = R$ so that Planck mass would be universal and equal to CP_2 mass $m_{CP_2} \sim 10^{-7} m_{Pl}$. For $m = Nm_{CP_2}$ one would have

$$\frac{\hbar_{gr}}{\hbar} = \sqrt{\frac{NRM_D}{\hbar v_0}} = \sqrt{\frac{NM_D}{m_{CP_2} v_0}} \ . \tag{2.16}$$

$N = 1$ would give $y(M_{CP_2}) \sim 10^{14.5}$ and $d \sim 10$ km. $n_{gr} = 10^{13}$ would require $d \sim 215$ km.

3. If one defines Planck mass $m_P = \hbar_{gr}/R$, one obtains for $N = 1$ $\hbar_{gr} = RM_D/v_0 = \hbar M_D/m_{CP_2}$. Now the condition does not say anything about m and expresses $\hbar_{gr}$ in terms of M_D. This option is perhaps the most elegant one.

There is still one objection. The cyclotron frequencies appearing in the applications are for elementary particles. The above formulas cannot be applied unless one replaces single charged particle with a BE condensate consisting of N particles with same charge ratio Ze/m so that cyclotron frequencies are unaffected and cyclotron transitions correspond to cyclotron phase transitions. This has been indeed assumed in the applications. In particular, fermionic ions are replaced with their Cooper pairs. The condition for having multiple pf Planck mass $m_P = \hbar_0/R$ is that Nm is K-multiple of $m_p = \hbar_0/R = \hbar/6R \sim 1.3\times 10^{14} m_p$. If there is one dark cyclotron proton per atomic volume Angstrom^3, this correspond volume $d = 5\mu$m, which defines cell size scale so that K could correspond to cell number.

2.2.6 Estimate of G/G_N from the delocalization at magnetic flux tubes

The following argument is for a situation in which the mass m corresponds to the mass of ion. By Equivalence Principle m however disappears from the formulas involving gravitational interaction of Earth, and cyclotron frequencies remain invariant for cyclotron BE condensate. Therefore the formulas apply for the BE condensate ions with total mass equal to a multiple of Planck mass $m_P = \hbar_0/R$.

The de-localization length of dark matter wave functions in the gravitational field is much longer than for ordinary value of Planck constant: essentially the height to which particle can rise with given initial velocity V_0 in the gravitational field with gravitational constant G. This would conform with the idea that dark particles are delocalized at the flux tubes in the scale of cyclotron wave-length.

The condition that the height h for the orbit equals to cyclotron wavelength gives an estimate for G_N/G. One can estimate the height $h = R - R_E$ from energy conservation assuming that particle has initial vertical velocity V_0 at the surface of Earth and cyclotron wavelength λ_c:

$$\frac{V_0^2}{2} = \frac{G}{G_N}\left[\frac{GM}{R_E} - \frac{GM}{R}\right] \ ,$$

$$h = \lambda_c = \frac{1}{f_c} = \frac{2\pi m}{neB} \ .$$

One obtains an estimate for G/G_N as

$$\frac{G}{G_N} = V_0^2\frac{(R_E+h)R_E}{r_S h} \ , \qquad R = R_E + h \ ,$$

$$h = \frac{\lambda_c}{n} = \frac{1}{nf_c} = \frac{2\pi m}{neB} \ . \tag{2.17}$$

This gives

$$\frac{G}{G_N} = nV_0^2 \times \frac{R_E(R_E+\frac{\lambda_c}{n})}{r_S\lambda_c} = nV_0^2 \times \frac{R_E(R_E+\frac{2\pi eB}{neBm})}{r_S} \times \frac{eB}{2\pi m} \ . \tag{2.18}$$

The condition that value of G/G_N is constant quantizes the value of V_0. For small value of h one has $V_0^2 n \simeq constant$. For $R_E \sim \lambda_c$ and nV_0^2 is of order unity, the order of magnitude would be $G/G_N \sim R_E/r_S \sim 7 \times 10^8$.

2.3 What can one say about the detailed anatomy of the MB?

The details of the anatomy of the MB have remained rather fuzzy hitherto. The following is an attempt to formulate more explicitly and coherently the earlier ideas scattered in books and articles about TGD. There are several empirical facts and theoretical constraints that one can use.

1. There is the notion of dark DNA as dark nuclei consisting of sequences of dark protons. The notion of dark nucleus is central concept in TGD based model of "cold fusion" [47]. Dark proton sequences are parallel with and in the vicinity of ordinary DNA strands and ordinary codons and dark proton triplets representing them [39] are paired.

2. Pollack effect [4] [38] for water is assumed to generate dark DNA. Part of protons go to the flux tube and negative charge is generated in ordinary matter and ends to negative charge of phosphates associated with the ordinary DNA nucleotides. Ordinary DNA would pair with dark DNA serving as predecessor and controller of ordinary DNA. Also RNA, amino-acids, and tRNA would have dark predecessors and similar pairing would occur.

3. Experiments of Peter Gariaev et al - in particular the discovery of phantom DNA [10] - and of Montagnier [7] [?] provide further valuable information.

Consider now what MB could look like.

1. MB has two parts. The small part has size of the physical system consisting of ordinary matter plus parts with relatively small h_{eff} assignable to structures such as valence bonds. The flux tubes of this part of MB connect parts of the system to a network and tensor network is an excellent mathematical model for what is involved. Flux tubes serve as topological correlates for entanglement and even prerequisites for it.

   In living matter one can imagine that the basic units of ordinary matter - say cells - are organized at parallel flux tubes. For $B_{end} = .2$ Gauss, which seems to define an especially important endogenous magnetic field, the radius r_B is of cell size. The value of proton cyclotron frequency is 300 Hz in this case and happens to correspond to the rotation frequency of the "shaft" of the ATPase as power generator.

   60 Hz frequency was found to lead to a transformation of cancer cells to ordinary ones and this suggests that cyclotron frequency for $B = B_{end}/5$ is involved. The flux tubes would contain 5 cells in their cross section and one can argue that dark proton quantum coherence at gravitational flux tubes with this thickness could give rise coherence in 5-cell length scale and lead to the cure of cancer.

2. The large part of MB - with size of the order Earth radius for $f_c = 60$ Hz corresponds to long flux tubes with large effective Planck constant $h_{gr}/h_0 = n$. Effective value of Planck constant is indeed in question since n_{gr} is the number sheets of the space-time surface as M^4 covering and Planck constant has value h_0 (rather than $h = 6h_0$) at each sheet of the covering. At QFT limit sheets are effectively replaced with single one, and one must allow the "real" Planck constant to have non-standard values.

   The cyclotron energies are scaled up by $h_{eff}/h_0 = n_{gr}$ and whatever the detailed anatomy of MB is this must be understood. Effectively one has n_{gr} photons with ordinary cyclotron energy and their energies sum up. This can be understood if the flux tubes define n_{gr}-fold coverings of M^4.

3. $h_{gr} = n_{gr}h_0$ correspond to quantum coherence in very long length scales whereas in the scale of organism the value of n is relatively small. The simplest idea is that n_{gr} disjoint flux tubes with small value of n and with given thickness determined by flux quantization coming from the living system combine to form single n_{gr}-sheeted flux tube with length given by $L_c = \lambda_c = 2\pi m/ZeB$ having no dependence on h_{eff}.

   This would be like a large number of cables combining a single cable. The threads of the cable would be now on top of each other in CP_2 direction! A rather exotic space savings! This would combine the sensory information coming from the separate flux tubes to a single super-cable and make the control of the system easy. Central nervous system would have spinal chord as an analogous unit both geometrically and functionally albeit in totally different scale. One of the first proposals was that MB provides an almost topographic representation of the biological body [19].

   One can estimate the volume of the region with coherence forced by quantum gravitational coherence as $V_{gr} = n_{gr}V(unit)$, where $V(unit)$ is the volume of the basic unit presumably determined by flux tube radius. If $V(unit)$ equals to volume a^3 of cube with side a, V_{gr} corresponds to a cube with side $a_{gr} = n_{gr}^{1/3}a$.

   The assumption that the energies of EEG photons in alpha band with $f = 10$ Hz correspond to ordinary photons at the lower end of the bio-photon spectrum having frequency 400 THz gives n_{gr} as $n_{gr} = 4 \times 10^{13}$. For $n_{gr} = 4 \times 1013$ and $a = 5$ μ m giving lower bound for the volume of neuron one would have $a_{gr} = 0.2$ m, roughly the size scale of brain.

4. The natural interpretation of the super-cables is as gravitational flux tubes. The gravitational flux associated with the ordinary flux tubes would combine to the dark gravitational flux tubes. This combination could take place repeatedly. Could the parameter M_D in $h_{gr} = n_{gr}h_0$ correspond to the portion of the Earth's gravitational flux flowing along these flux tubes? The sum of the masses

M_D should over values of field strengths and charged particle masses should give the total mass M_E of Earth if the guess is correct.

One must of course be extremely cautious in interpretations. For instance, flux tubes carrying Kähler charge the flux tubes should be closed and give rise to a kind of Dirac monopole like structure with return flux. This would mean that gravitational flux returns back, possibly along different space-time sheets. But the flux lines are closed also for the ordinary magnetic fields. Can this really be consistent with the Newtonian view about gravitation in which gravitational flux flows to infinity? The answer is far from obvious: the many-sheeted space-time in which space-time sheets are glued along the boundaries would that part of the flux can return and part goes to larger space-time sheets and in principle there is no largest space-time sheet so that one would obtain effectively monopoles.

5. An entire fractal hierarchy of magnetic field strengths is predicted. A good guess is that field strengths are given by p-adic length scale hypothesis, that is have scales given by $B(k) \propto 1/L(k)^2$, where $L(k) \propto 2^{k/2}$ is the p-adic length scale assignable to $p \simeq 2^k$. This would mean hierarchy of flux tubes with radii $L(k)$ and at each level the combination to super-cables representing gravitational flux tubes would take place.

   One has $M_D \propto v_0/\sqrt{B} \propto v_0 2^{k/2}$. For a fixed value of v_0, the sum can converge only if the number of p-adic length scales involved is finite. The radius R_E of Earth certainly gives this kind of upper bound and corresponds to a rather modest value of k ($L(151)$ correspond to 10 nm) . Also v_0 can depend on p-adic length scale. The sizes of living organisms give a more stringent upper bound on k.

2.4 Water memory and homeopathy

There is a lot of support about the representation of water memory as extremely low frequencies (ELF) of radiation associated with water [5, 6]. These ELF frequencies can be stored electronically and they produce the same effects as the bio-active chemical, whose presence induced these frequencies in water. At the age of IT the idea about the existence of representations of bio-active molecules as frequency patterns able to induce the biological effects of molecules without the presence of molecules should not raise grave objections. For instance, brain generates this kind of representations by entrainment to external frequencies and water might play a crucial role also here. Few years ago HIV Nobelist Montagnier did experiments giving support for water memory and the procedure involved a part very similar to that used in preparing homeopathic remedies [7, 36].

The description of water memory in TGD Universe would look like follows.

1. In TGD framework these frequencies would correspond to cyclotron frequencies assignable to MBs of molecules, and immune system is proposed to have emerged from the ability of water to mimic the MBs of invader molecules and learning to recognize them [17] by resonant coupling at these frequencies.

   This would take place via entrainment made possible by the variation of the thickness of the flux tube inducing variation of the cyclotron frequency. In entrainment the cyclotron frequency of the flux tube would co-incide with the external frequency. MB having flux tubes with modified thickness would be able to produce cyclotron radiation at the these frequencies and couple to the invader molecule resonantly. The coupling would involve also topological part as reconnection of flux tubes with same thickness and carrying same charged particles to make resonance possible.

   One can visualize living systems as systems having magnetic tentacles consisting of U-shaped flux tubes forming thus locally pairs of flux tube tubes and searching for flux similar flux tubes of other systems, in particular bio-active molecules. The recognition of invader molecules is a crucial part of immune systems and this mechanism would be an essential part of immune action besides cyclotron resonance.

2. In TGD universe water is very special substance in that it contains both ordinary water and its dark variant. What makes it dark is that dark magnetic flux tubes representing long hydrogen bonds are present for some portion of water [56] (see `http://tinyurl.com/y8fvwbp9`): the length of bonds scales as n or perhaps even n^2. The presence of these flux tubes makes any liquid phase a network like structure, and one ends up with a model explaining an anomaly of thermodynamics of liquids at criticality known already in Maxwell's time. This leads to a model explaining the numerous anomalies of water in terms of the dark matter.

   For instance, the dark part of water with non-standard Planck constant transforms to ordinary water in freezing. As a consequence, a large amount of energy is liberated. This explains why water has anomalously large latent heat of fusion. One can also understand why the volume of water increases in freezing and decreases in heating in the interval 0-4 °C. The anomalies of water are largest at physiological temperature $T_{phys} \sim 37$ °C suggesting that the dark portion of water is largest at T_{phys}. Dark fraction of water would be essential for life.

3. Pollack effect [4] (see `http://tinyurl.com/oyhstc2`) requiring feed of energy - as IR radiation for instance - generates so called exclusion zones (EZs), which are negatively charged regions. A fraction of protons from water must go somewhere and the TGD inspired proposal [38] (see `http://tinyurl.com/gwasd8o`) is that the protons transform to dark protons at magnetic flux tubes. The dark variants of particles quite generally have higher energies than ordinary ones and energy feed provides the needed metabolic energy go make the protons dark. In the case of homeopathy and water memory mechanical agitation creates provides the metabolic energy and would generate EZs accompanied by dark proton sequencies at flux tubes [17].

4. The MB of water would be also a key central part of MB of the living system acting as intentional agent receiving sensory input from biological body and controlling it. Biochemistry would be kind of shadow dynamics. The ions provided by the living system would reside at the flux tubes of MB provided by water and as found the lengths of flux tubes and also the value of $h_{eff} = h_{gr}$ at the would distinguish between different ions. The gravitational flux tubes formed by combination of n_{gr} ordinary flux tubes to n_{gr} flux tubes with the same M^4 projection defining a covering of M^4 would define the large part of MB serving as intentional agent and communications would occur at cyclotron frequencies.

   Cell membranes would produce what I call generalized Josephson radiation, which would couple resonantly to cyclotron Bose-Einstein condensates at the flux tubes. Nerve pulse patterns would induce frequency modulation allowing to code sensory input represented by them and send it to MB which in turn could send control signals through genome [23, 15, 16, 26].

   MB would be the seat of primary form of genetic code. Dark protons sequences at flux tubes representing genetic code [39] and the analogs of the other basic biomolecules are realized in water.

2.5 What the view about magnetic body could mean at the level of DNA and other basic bio-molecules?

A more precise vision about the anatomy of MB leads to a flux of ideas and questions. Flux tubes from identical basic units (cells, DNA, identical proteins, etc) combine to form single many-sheeted flux tube so that the incoming flux tubes have same M^4 projection being on top of each other in CP_2 direction. This super cable is like umbilical chord! The structures form a Bose-Einstein condensate in abstract topological sense.

This opens fascinating possibilities for understanding of dark DNA..

1. Cells have identical DNAs. Earlier I have assumed that magnetic flux sheets go through DNA in transversal direction and that dark DNA in some sense is sequence of dark proton triplets associated

with flux tubes. Furthermore, DNA transcription requires that there are transversal flux tubes emerging from codons or perhaps even from nucleotides as flux tubes inside codon flux tube.

How to combine these views together with new view about combination of the DNAs flux tube to larger superstructure, one DNA from each cell in structure?

2. For single DNA each codon would correspond to 3-proton units organized linearly into a sequence. Each 3-proton unit must have a flux tube transversal flux to DNA strand and located at 2-D sheet. This brings in mind the structure of spine as anatomical and neurobiological analogy. This suggests that dark DNA codons formed by 3-proton units should be associated with these horizontal flux tubes in 2-D locally planar surface going through DNA.

3. These structures from $n_{gr} = h_{gr}/h_0 = h_{eff}/h_0$ separate cells should combine to single n_{gr}-sheeted gravitational flux tube with sheets on top of each other with same M^4 projection. This would be dark DNA at the level of MB. It would seem that given codon of each DNA must contribute a dark proton triplet so that there would be n_{gr} dark proton triplets at given flux tube which is however very long. The size scale - that is the length of the flux tube - is that of Earth typically and fixed by the cyclotron wave length λ_c.

   This would give a concrete topological meaning to quantum quantum coherence at the level of MB of bio-system. Also a view about how lower level conscious entities integrate to larger ones: one can imagine entire fractal hierarchy of structures integrating to larger structures integrating... In particular, altered states of consciousness could correspond to this kind of temporary integrations to higher level structures. Same should apply to other basic biological structures: cells, proteins, RNA, tRA. Dark realization of the genetic code predicts the dark variants of these biomolecules.

   This picture conforms with adelic physics [48, 49] in which n_{gr} corresponds to the dimension of extension of rationals: the larger the value of n_{gr}, the higher the algebraic complexity and level of conscious intelligence.

4. Where are the dark protons and various dark ions at dark flux tubes? Along entire long flux tubes with length of order cyclotron wavelength for given charged particle? Or inside the organism?

   The model of dark DNA allows only the latter option. They must reside at the short portions of the magnetic flux tubes inside organism. For instance, the dark protons of dark DNA are associated with flux tubes parallel and in immediate vicinity of ordinary DNA strand and codon and dark codon a paired like codon and its conjugate in ordinary DNA.

   What makes these particles dark is that they are controlled by the gravitational flux tube and form a non-local quantum coherent unit containing n_{gr} particles.

This raises a long series of questions and fantastic challenges for visual imagination.

1. How do DNA and its conjugate relate at this level: do DNA and conjugate correspond to single closed long flux tube forming part of the "umbical chord" far from biological body?

2. What replication of DNA could mean topologically at the level of this super-DNA? What about description of transcription and translation at these super-levels. Are the ordinary replication, etc.. induced from this super level as mere shadow processes: this would explain their mysterious coherence?

3. What sexual reproduction and associated recombination of chromosomes could mean at super level? What does the growth of organisms mean at super level? Addition of new sheets to super DNA and its variants so that n_{gr} defined as the number of basic units grows and organism becomes more and more quantum intelligent?

3 Two applications of the model of magnetic body

In the following the model of MB is applied to explain the findings of Montagnier [7]. and of Li and Heroux [8].

3.1 Interpretation of the experiments of Montagnier et al

One can make the model of MB more detailed by applying it the experiments of Montagnier et al [7] discussed earlier from TGD viewpoint at [36]. I have developed in collaboration with Peter Gariaev and analogous model analogous observations [33].

Consider first a very rough sketch of the experiments.

1. A fragment of DNA was amplified by PCR. It was diluted to say 10^{-6} by adding pure water and found to generate EMS at ULF frequency range .5-3 kHz. Call this sample A.

2. Dilution was put in mu-mental container, which does not allow ULF radiation to get out. In its vicinity another tube, call it B, containing pure water was placed. The water content of each tube was filtered through 450 nm and 20 nm filters. Filtering does not allow particles with size smaller than 20 nm to go through. The samples were diluted to 10^{-15} by adding pure water. During each dilution a mechanical agitation of water by generating a vortex was performed.

3. Copper solenoid producing 7 Hz current was added around the samples. Eventually EMS was found in *both* A and B. In B there was primer, DNA polymerase, and free nucleotides but not the template complementary strand as in ordinary polymerase chain reaction.

The TGD interpretation for what happened in sample A would be following.

1. As explained ordinary DNA strand is paired with dark DNA strand for which dark proton triplets at flux tube parallel to ordinary DNA strand represent the codons [39]. This is analog of pairing of DNA strand and its conjugate.

2. The cyclotron transitions of dark protons (possibly also those changing the direction of spin) generate the ULF radiation as classical em fields accompanied dark photons transformed to ordinary photons. The energies of dark photons are given by $E = h_{gr}f$ and should be above thermal energy at physiological temperatures. The transformations of dark photons to ordinary photons give rise to bio-photons with energies in visible and UV, and possibly also below this range [28, 29].

3. Dilution eliminated ordinary DNA from A but left some dark DNA strands to the water. This is nothing but the phantom DNA phenomenon [10] discovered by Gariaev and collaborators [35]. In case of water memory one has phantoms of bio-active molecules [17]. I have applied TGD also to other experimental findings and ideas by Gariaev et al [9, 11, 12]. In particular there are articles written in collaboration about the TGD realization of identification of DNA as hologram [32] and about DNA remote replication analogous to what happens in Montagnier's experiments [33].

4. The interpretation of the agitation carried out also in the preparation of homeopathic remedies is that it provided metabolic energy needed to generate large h_{eff} [17]. Quite generally, the energy of dark variant is larger than that of ordinary state: for instance, cyclotron energy is proportional to h_{eff} and atomic binding energies to $1/h_{eff}^2$ so that metabolic energy is needed.

   The analogy with ordinary DNA and the idea that DNA replication is a shadow of the replication of dark DNA suggests that dark DNA replicated and a population of dark DNA mimicking ordinary DNA was generated in the diluted water sample A. More generally, water would perform mimicry of bio-active molecule by using dark protons at its magnetic body to generate the cyclotron frequency spectrum of the molecule. An interesting possibility is that dark proton sequences - dark nuclei - perform this mimicry.

5. The general model suggests that dark DNA generate transversal flux tubes at transversal sheets going through it. One could start by saying loosely that dark photon cyclotron radiation propagated along these flux tubes to the pure water sample, where there was flux tube receiving this radiation. But what the precise meaning of this statement could be, becomes more clear in the following.

What happened in the pure water sample?

1. Dark photon radiation at ULF frequencies should have caused the generation of dark DNA strands also in pure water sample. The water in the pure water sample mimicked the dark photon radiation by the basic homeopathic mechanism and generated dark DNA strands with transversal flux tubes at transversal flux sheets carrying magnetic fields corresponding to the cyclotron frequencies of dark DNA nucleotides.

2. Did a pairing of dark DNA in A and its conjugate in B analogous to the pairing of DNA and its conjugate take place? As in the ordinary DNA pairing the pairing would be favored by the minimization of interaction energy. The flux tubes connecting the members of the pair would be like stretched hydrogen bonds between DNA strand and its conjugate so that it would be very long, of order cyclotron wavelength of proton in the magnetic field of flux tube. TGD indeed predicts that hydrogen bonds have length spectrum corresponding to various values of h_{eff} [56]. In this case however rather small values are involved. Now the values would be very large by $h_{eff} = h_{gr}$.

3. Did the transversal flux tubes attach to the dark DNA flux tubes directly and have length of the order of distance between samples? Or did the flux tubes from water sample combine to many-sheeted gravitational flux tube with length of the order of cyclotron frequency of proton? The latter option is favored.

   The dark photons assignable to the frequency range .5-3 kHz should have energies above thermal energy at physiological temperature in order to have physiological effects. This requires $h_{eff} \geq 6 \times 10^{10}$: $f = .5$ kHz would correspond to a thermal energy of photon for most probable wavelength about 10 μm.

   A reasonable estimate for the length of the flux tubes involved comes from the cyclotron wavelength of proton. The cyclotron frequency $.5 - 3.0$ kHz for proton requires magnetic field about $.3 - 2.0$ Gauss somewhat stronger than $B_{end} = .2$ Gauss ($B_E = .5$ Gauss). The cyclotron wavelength would be in the range .1-.6 Mm (Earth radius is 6.4 Mm) so that the analogs of hydrogen bonds would be very long!

4. The generation of ordinary DNA strands in this sample would have been by the analog of DNA - dark DNA pairing that should occur in standard biology. The DNA fragment in pure water sample was reproduced as if the complementary strand would have been present.

5. As already explained, the dark protons are at single sheeted ordinary flux tubes accompanying DNA. n_{gr} flux tubes from various positions would combine n_{gr}-sheeted flux tube with same M^4 projection as ordinary flux tube: they were on top of each other in $M^4 \times CP_2$. Why the protons deserved to be called dark is that the proton sequences at n_{gr} separate flux tubes form single quantum coherent many-proton states somewhat analogous to Bose-Einstein condensate. Cyclotron energy is therefore naturally n_{gr} times the cyclotron energy of single state. This is essentially quantum non-locality made possible by the locality at the gravitational flux tube quantum controlling the system. Also dark photons having n_{gr}-fold energy non-local many-photon states with one photon at each flux tube with same energy and momentum.

6. Also 7 Hz frequency is necessary for the effect to occur. A natural guess is that this frequency is related to Schumann resonances (see `http://tinyurl.com/cv8z9vs`), which are associated with collective oscillations in the Earth's magnetic field in the cavity bounded by Earth's surface and

ionosphere in which em waves cannot propagate. Schumann resonances dominate the frequency spectrum from 3 Hz to 60 Hz - a considerable part of EEG - and there are distinct peaks at frequencies resonance 7.81, 14.3, 20.8, 27.3 and 33.8 Hz.

In the linear model Schumann resonances treating atmosphere as vacuum Schumann resonance frequencies $f_n = \sqrt{n(n+1)}c/2\pi R_E$ are determined by the geometry alone with lowest resonance at 7.41 Hz for $R_E = 6371$ km. The finite conductivity of atmosphere lowers the propagation velocity of light and the frequencies are reduced. This can bring the resonance frequency 7.81 Hz nearer to 7 Hz, and there is of course also the continuum besides the resonance peaks.

In TGD picture the quasi-continuum would relate to the many-sheetedness of the space-time surface making it possible for light to propagate along large number of flux tube paths so that the effective light-velocity would vary. A more precise model give also a resonance at 4.11 Hz. This resonance frequency however varies due to the several factors.

The interpretation of Schumann resonances 7.81, 14.3, 20.8, 27.3 and 33.8 Hz and higher resonances as resonance frequencies of EEG is highly attractive. At higher frequencies the resonances appear approximately with 6.5 Hz intervals. Next resonance would be at 40.3 Hz, which is the familiar thalamo-cortical resonance frequency to which consciousness was once assigned. 8:th partial wave has resonance frequency 60 Hz which happens to be the frequency appearing in the experiments of Li and Heroux. 4 Hz frequency in turn is theta resonance frequency in EEG.

This supports the view that water entrains to Schumann frequencies by tuning to cyclotron frequencies by varying the thickness of flux tubes of its MB so that the coupling of living matter to the oscillations of Earth's magnetic field would play fundamental role in biology and neuroscience. The testable prediction is the correlation of EEG with the local Schumann resonance spectrum of Earth independent of individual.

Dr. Phil Callahan [13, 14] claims on basis of intensive experimental work that there is a tendency of political strifes and wars to concentrate on regions where Schumann resonances are weak. In the proposed picture this would not be surprising. The reduction in the level of consciousness would imply strifes and wars at the level of society and cancer at the level of cell community.

3.2 Cancer as a disease of the MB of water?

How the irradiation of the cancer cell population with 60 Hz oscillating magnetic field with extremely small intensity in above 25 nT could lead to the reduction of the chromosome number of mitochondria and return of cancer cells to a normal state?

One should locate the problem. Is the problem at the level of cell membrane, mitochondria, or ATPase as Josephson junction, or possibly at the level of MB of water? Could cancer - and perhaps many other diseases - be diseases of the MB of water? This option is certainly the simplest one since one can forget entire chemistry of cells apart from the presence of charged particles.

What the problem is? Cancer rather obviously means a loss of coherence at the level of ordinary bio-matter forced by quantum coherence at the level of MB. This suggests that the quantum coherence of MB is for some reason lost in cancer. The coherent behavior of cell groups consisting of few cells is lost and cells behave like individuals knowing nothing about each other's presence.

In the recent case one can consider protons in magnetic field $B = B_{end}/5$ with 5 times larger flux tube area increasing the area of quantum control by factor 5: 5 cells instead of 1 cells roughly. One can also consider magnetic field $B = 6B_{end}/5$ with Li^+ ions having mass number $A = 6$. Now the length rather than radius of flux tubes would be scaled up by factor 5. Li^+ ions are indeed applied in manic depressive disease and schizophrenia but the mechanism for healing is unknown. TGD proposal is that their presence generates cyclotron radiation needed to have communications with the layers of MB responsible for the control of axons for instance known to suffer from inflammation [43, 40].

Let us consider a more detailed model.

1. Assume that MB, whose anatomy has been described above controls bio-matter. The radius of flux tube defines the size scale of coherent regions and the quantum coherence of gravitational part of MB can force coherence in this regions. For cancer cells the radius of this regions is cell radius and flux tubes with this thickness form the basic structural units. For $B_{end} = .2$ Gauss the radius $R_B \simeq 3.2$ μm of the flux tube is indeed of the order of cell radius. The cyclotron frequency for proton is 300 Hz and this conforms with the idea that it defines the rotation frequency of the shaft of the power generator defined by ATPase.

2. The coherent behavior requires the presence of also higher levels in the p-adic length scale hierarchy. 60 Hz frequency corresponds to cyclotron frequency for dark protons in $B = B_{end}/5$. This does not actually correspond to power of 2 but I have proposed that also powers of other p-adic length scales for small primes could be important and their is evidence for $p = 3$ [21]. Now one would have $p = 5$. Note that the roots $\sqrt{p}$, $p \in \{2, 3, 5\}$ are in a central role in the geometry of Platonic solids (the geometry of icosahedron is in central role in one TGD based model of genetic code based on the notion of bio-harmony [37, 57].

   Since flux tube with 5-fold area contains 5 cells in transversal cross section, this suggests that in cancer the coherent behavior at 5-cell level is lost. It might be that the thickness of these flux tubes has for some reason changed so that they are out of synch or that they are missing altogether.

3. The above summarized experiments of Russians [3] show that the physical properties of water change by irradiation with extremely weak magnetic fields at frequencies at various frequencies. That the properties of water change would be due to the control action by MB of the water. The effect depends very little on the strength of the field and this conforms with the entrainment hypothesis meaning that flux tubes tune their thickness to achieve resonance like radio set.

   TGD interpretation is in terms of water memory. In TGD water memory is represented as cyclotron frequencies associated with the flux tubes of MB of water, its body parts characterized by various frequencies and the body parts, flux tubes, can thicken in which case the frequency is reduced and vice versa. Even new body parts can emerge and it is possible genetic code codes for them (in fact dark genome assignable to protonic flux tubes parallel to DNA would be the fundamental code). MB entrains to external frequencies by varying the thickness of its flux tubes and can respond to and represent them as cyclotron frequencies.

   The healing of cancer cells by 60 Hz radiation could bring to the MB of cancer cells the protonic flux tubes with $f_c = 60$ Hz. The communications to the big MB of the cell would be restored and MB could take care of the cell.

4. The above discussed connection with Schumann resonances suggests that all Schumann resonances are fundamental for biology and the MB of water entrains to them. The 8:th Schumann resonance is indeed 60 Hz.

This could have rather far reaching implications.

1. Also the homeopathic treatment of water [17] is explained in terms of the generation of flux tubes as body parts of MB of water having the cyclotron frequencies of the molecules involved in the treatment. These molecules represent at least the ELF part of the cyclotron energy spectrum of molecules and can therefore couple resonantly to them so that MB can detect the invader molecules and molecules of immune system can catch them and make them harmless.

   The mechanical agitation of water would provide the metabolic energy needed to general the needed large values of h_{eff}. Quite generally, metabolic energy feed is needed in order to generate MB containing the dark matter. At least the ELF spectrum of homeopathic remedy has been found to generate the same effective as the original molecule, which also demonstrates that it is ELF spectrum, which is responsible for bio-activity. This does not make sense in standard quantum theory since the photon energies are much below the thermal energy at the physiological temperatures. The large value of h_{gr} saves the situation.

2. If this picture is correct, homeopathic remedies could be also generated by irradiation of water using the ELF frequency spectrum characterizing the substance considered.

3. In the case of cancer the irradiation of water with 60 Hz frequency could generate the required body parts of MB or get existing body parts in synch and induce a healing of cancer. Cancer would be very probably only very special case if MB is the intentional agent controlling biological body in various scales and bio-chemistry is a kind of shadow dynamics as suggested. One can even imagine a medicine completely free of the side effects of various chemical medicines [17]. If even genetic diseases are basically diseases of MB, they might be healed homeopathically. An interesting question is whether water is a kind of universal emulator of various molecules able to very quickly modify its MB.

References

[1] Preskill J et al. Holographic quantum error-correcting codes: Toy models for the bulk/boundary correspondence. Available at: `http://arxiv.org/pdf/1503.06237.pdf`, 2015.

[2] Nottale L Da Rocha D. Gravitational Structure Formation in Scale Relativity. Available at: `http://arxiv.org/abs/astro-ph/0310036`, 2003.

[3] Effect of weak magnetic fields on the properties of water and ice. *Soviet Physics Journal.* Available at:`http://tinyurl.com/yd7kqnzg`, 31(5), 1988.

[4] The Fourth Phase of Water : Dr. Gerald Pollack at TEDxGuelphU. Available at: `https://www.youtube.com/watch?v=i-T7tCMUDXU`, 2014.

[5] Benveniste J et al. Human basophil degranulation triggered by very dilute antiserum against IgE. *Nature*, 333:816–818, 1988.

[6] Benveniste J et al. Transatlantic transfer of digitized antigen signal by telephone link. *J Allergy and Clinical Immunology.* Available at: `http://www.digibio-.com/`, 99:175, 1989.

[7] Montagnier L et al. DNA waves and water. Available at: `http://arxiv.org/abs/1012.5166`, 2010.

[8] Li Y and Heroux P. *Electromagnetic Biology ad Mecicine.* Available at: `http://tinyurl.com/y9lv47qp`, 33(4), 2014.

[9] Gariaev P et al. *The DNA-wave biocomputer*, volume 10. CHAOS, 2001.

[10] Gariaev PP et al. Holographic Associative Memory of Biological Systems. *Proceedings SPIE - The International Soc for Opt Eng . Optical Memory and Neural Networks*, pages 280–291, 1991.

[11] Gariaev PP et al. The spectroscopy of bio-photons in non-local genetic regulation. *J Non-Locality and Remote Mental Interactions.* Available at: `http://www.emergentmind.org/gariaevI3.htm`, (3), 2002.

[12] Tovmash AV Gariaev PP, Tertishni GG. Experimental investigation in vitro of holographic mapping and holographic transposition of DNA in conjuction with the information pool encircling DNA. *New Medical Technologies*, 9:42–53, 2007.

[13] Earth's magnetic field regions of weakness correlated to sites of political unrest and war: the paradigm quaking measurements of professor Phil Callahan. Available at: `http://www.acacialand.com/Callahan.html`.

[14] Dr. Phil Callahan on Power of Paramagnetism. *Nexus.* Available at: http://www.nexusmagazine.com, February 2003.

[15] Pitkänen M. Dark Matter Hierarchy and Hierarchy of EEGs. In *TGD and EEG.* Online book. Available at: http://www.tgdtheory.fi/public_html/tgdeeg/tgdeeg.html#eegdark, 2006.

[16] Pitkänen M. DNA as Topological Quantum Computer. In *Genes and Memes.* Online book. Available at: http://www.tgdtheory.fi/public_html/genememe/genememe.html#dnatqc, 2006.

[17] Pitkänen M. Homeopathy in Many-Sheeted Space-Time. In *Bio-Systems as Conscious Holograms.* Online book. Available at: http://www.tgdtheory.fi/public_html/hologram/hologram.html#homeoc, 2006.

[18] Pitkänen M. Magnetic Sensory Canvas Hypothesis. In *TGD and EEG.* Online book. Available at: http://www.tgdtheory.fi/public_html/tgdeeg/tgdeeg.html#mec, 2006.

[19] Pitkänen M. *Magnetospheric Consciousness.* Online book. Available at: http://www.tgdtheory.fi/public_html/magnconsc/magnconsc.html, 2006.

[20] Pitkänen M. Negentropy Maximization Principle. In *TGD Inspired Theory of Consciousness.* Online book. Available at: http://www.tgdtheory.fi/public_html/tgdconsc/tgdconsc.html#nmpc, 2006.

[21] Pitkänen M. p-Adic Physics as Physics of Cognition and Intention. In *TGD Inspired Theory of Consciousness.* Online book. Available at: http://www.tgdtheory.fi/public_html/tgdconsc/tgdconsc.html#cognic, 2006.

[22] Pitkänen M. Quantum Model for Hearing. In *TGD and EEG.* Online book. Available at: http://www.tgdtheory.fi/public_html/tgdeeg/tgdeeg.html#hearing, 2006.

[23] Pitkänen M. Quantum Model for Nerve Pulse. In *TGD and EEG.* Online book. Available at: http://www.tgdtheory.fi/public_html/tgdeeg/tgdeeg.html#pulse, 2006.

[24] Pitkänen M. *TGD and EEG.* Online book. Available at: http://www.tgdtheory.fi/public_html/tgdeeg/tgdeeg.html, 2006.

[25] Pitkänen M. *TGD Inspired Theory of Consciousness.* Online book. Available at: http://www.tgdtheory.fi/public_html/tgdconsc/tgdconsc.html, 2006.

[26] Pitkänen M. Three new physics realizations of the genetic code and the role of dark matter in bio-systems. In *Genes and Memes.* Online book. Available at: http://www.tgdtheory.fi/public_html/genememe/genememe.html#dnatqccodes, 2006.

[27] Pitkänen M. Quantum Mind, Magnetic Body, and Biological Body. In *TGD based view about living matter and remote mental interactions.* Online book. Available at: http://www.tgdtheory.fi/public_html/tgdlian/tgdlian.html#lianPB, 2012.

[28] Pitkänen M. Are dark photons behind biophotons. In *TGD based view about living matter and remote mental interactions.* Online book. Available at: http://www.tgdtheory.fi/public_html/tgdlian/tgdlian.html#biophotonslian, 2013.

[29] Pitkänen M. Comments on the recent experiments by the group of Michael Persinger. In *TGD based view about living matter and remote mental interactions.* Online book. Available at: http://www.tgdtheory.fi/public_html/tgdlian/tgdlian.html#persconsc, 2013.

[30] Pitkänen M. Criticality and dark matter. In *Hyper-finite Factors and Dark Matter Hierarchy.* Online book. Available at: http://www.tgdtheory.fi/public_html/neuplanck/neuplanck.html#qcritdark, 2014.

[31] Pitkänen M. Quantum gravity, dark matter, and prebiotic evolution. In *Genes and Memes.* Online book. Available at: http://www.tgdtheory.fi/public_html/genememe/genememe.html#hgrprebio, 2014.

[32] Pitkänen M Gariaev P. Model for the Findings about Hologram Generating Properties of DNA. In *Genes and Memes.* Online book. Available at: http://www.tgdtheory.fi/public_html/genememe/genememe.html#dnahologram, 2011.

[33] Gariaev P Pitkänen M. Quantum Model for Remote Replication. In *Genes and Memes.* Online book. Available at: http://www.tgdtheory.fi/public_html/genememe/genememe.html#remotereplication, 2011.

[34] Pitkänen M. Quantum Astrophysics. In *Physics in Many-Sheeted Space-Time.* Online book. Available at: http://www.tgdtheory.fi/public_html/tgdclass/tgdclass.html#qastro, 2006.

[35] Pitkänen M. Wormhole Magnetic Fields. In *Quantum Hardware of Living Matter.* Online book. Available at: http://www.tgdtheory.fi/public_html/bioware/bioware.html#wormc, 2006.

[36] Pitkänen M. DNA Waves and Water . Available at: http://tgdtheory.fi/public_html/articles/mont.pdf, 2011.

[37] Pitkänen M. Geometric theory of harmony. Available at: http://tgdtheory.fi/public_html/articles/harmonytheory.pdf, 2014.

[38] Pitkänen M. Pollack's Findings about Fourth phase of Water : TGD View. Available at: http://tgdtheory.fi/public_html/articles/PollackYoutube.pdf, 2014.

[39] Pitkänen M. About Physical Representations of Genetic Code in Terms of Dark Nuclear Strings. Available at: http://tgdtheory.fi/public_html/articles/genecodemodels.pdf, 2016.

[40] Pitkänen M. Are lithium, phosphate, and Posner molecule fundamental for quantum biology? Available at: http://tgdtheory.fi/public_html/articles/fisherP.pdf, 2016.

[41] Pitkänen M. Holography and Quantum Error Correcting Codes: TGD View. Available at: http://tgdtheory.fi/public_html/articles/tensornet.pdf, 2016.

[42] Pitkänen M. Hydrinos again. Available at: http://tgdtheory.fi/public_html/articles/Millsagain.pdf, 2016.

[43] Pitkänen M. Lithium and Brain . Available at: http://tgdtheory.fi/public_html/articles/lithiumbrain.pdf, 2016.

[44] Pitkänen M. p-Adicizable discrete variants of classical Lie groups and coset spaces in TGD framework. Available at: http://tgdtheory.fi/public_html/articles/padicgeom.pdf, 2016.

[45] Pitkänen M. Artificial Intelligence, Natural Intelligence, and TGD. Available at: http://tgdtheory.fi/public_html/articles/AITGD.pdf, 2017.

[46] Pitkänen M. Does valence bond theory relate to the hierarchy of Planck constants? Available at: http://tgdtheory.fi/public_html/articles/valenceheff.pdf, 2017.

[47] Pitkänen M. Cold fusion, low energy nuclear reactions, or dark nuclear synthesis? Available at: http://tgdtheory.fi/public_html/articles/krivit.pdf, 2017.

[48] Pitkänen M. Philosophy of Adelic Physics. In *Trends and Mathematical Methods in Interdisciplinary Mathematical Sciences*, pages 241–319. Springer.Available at: https://link.springer.com/chapter/10.1007/978-3-319-55612-3_11, 2017.

[49] Pitkänen M. Philosophy of Adelic Physics. Available at: http://tgdtheory.fi/public_html/articles/adelephysics.pdf, 2017.

[50] Pitkänen M. About dark variants of DNA, RNA, and amino-acids. Available at: http://tgdtheory.fi/public_html/articles/darkvariants.pdf, 2018.

[51] Pitkänen M. About the physical interpretation of the velocity parameter in the formula for the gravitational Planck constant . Available at: http://tgdtheory.fi/public_html/articles/vzero.pdf, 2018.

[52] Pitkänen M. Is the hierarchy of Planck constants behind the reported variation of Newton's constant? Available at: http://tgdtheory.fi/public_html/articles/variableG.pdf, 2018.

[53] Pitkänen M. Clustering of RNA polymerase molecules and Comorosan effect. Available at: http://tgdtheory.fi/public_html/articles/clusterRNA.pdf, 2018.

[54] Pitkänen M. Dark valence electrons and color vision. Available at: http://tgdtheory.fi/public_html/articles/colorvision.pdf, 2018.

[55] Pitkänen M. Emotions as sensory percepts about the state of magnetic body? Available at: http://tgdtheory.fi/public_html/articles/emotions.pdf, 2018.

[56] Pitkänen M. Maxwells lever rule and expansion of water in freezing: two poorly understood phenomena. Available at: http://tgdtheory.fi/public_html/articles/leverule.pdf, 2018.

[57] Pitkänen M. New results in the model of bio-harmony. Available at: http://tgdtheory.fi/public_html/articles/harmonynew.pdf, 2018.

[58] Pitkänen M. Sensory perception and motor action as time reversals of each other: a royal road to the understanding of other minds? Available at: http://tgdtheory.fi/public_html/articles/timemirror.pdf, 2018.

Made in the USA
Las Vegas, NV
10 July 2024